ADVANCES IN
TWIN AND SIB-PAIR ANALYSIS

ADVANCES IN TWIN AND SIB-PAIR ANALYSIS

Edited by

Tim D. Spector, Harold Snieder

and

Alex J. MacGregor

all at

Twin Research & Genetic Epidemiology Unit
St. Thomas' Hospital
London, UK

© 2000
Greenwich Medical Media Ltd.
137 Euston Road
London
NW1 2AA

ISBN 1 84110 004 8

First Published 2000

A catalogue record for this book is available from the British Library

Distributed worldwide by
Oxford University Press

Project Manager
Gavin Smith

Production and Design by
Saxon Graphics Limited, Derby

Printed in Great Britain

CONTENTS

CONTRIBUTORS

David B. Allison
Obesity Research Center
St. Luke's / Roosevelt Hospital
Columbia University College of Physicians and Surgeons
USA

Dorret I. Boomsma
Free University
Department of Biological Psychology
Amsterdam
The Netherlands

Changzhong Chen
Anhui MEIZHONG Institute for Biomedical Sciences and
Environmental Health
Anhui
China

Joe C. Christian
Department of Medical and Molecular Genetics
Indiana University School of Medicine
Indianapolis
Indiana
USA

Conor V. Dolan
University of Amsterdam
Department of Psychology
Amsterdam
The Netherlands

David Duffy
Queensland Institute of Medical Research
Bancroft Centre
Brisbane
Queensland
Australia

Myles S. Faith
Obesity Research Center
St. Luke's / Roosevelt Hospital
Columbia University College of Physicians and Surgeons
USA

John L. Hopper
The University of Melbourne
Department of Public Health and Community Medicine
Carlton, Victoria
Australia

Jaako Kaprio
Department of Public Health
University of Helsinki
Department of Mental Health and Alcohol Research
National Public Health Institute
Helsinki
Finland

Markku Koskenvuo
Department of Public Health
University of Turku
Turku
Finland

Cathryn M. Lewis
Division of Medical and Molecular Genetics
Guy's, King's & St. Thomas' School of Medicine
King's College
London
UK

Alex J. MacGregor
Twin Research & Genetic Epidemiology Unit
St Thomas' Hospital
London
UK

Nick Martin
Queensland Institute of Medical Research
Bancroft Centre
Brisbane
Queensland
Australia

Michael C. Neale
Department of Psychiatry
Virginia Commonwealth University
Virginia Institute for Psychiatric & Behavioral Genetics
Richmond, Virginia
USA

Tianhua Niu
Program for Population Genetics
Harvard School of Public Health
Boston, Massachusetts
USA

Kirsten Ohm Kyvik
The Danish Twin Register
Institute of Community Health
Odense University
Odense
Denmark

David I.W. Phillips
Metabolic Programming Group
MRC Environmental Epidemiology Unit
Southampton General Hospital
Southampton
UK

Andrew Pickles
School of Epidemiology and Health Science
University of Manchester
Manchester
UK

John Rogus
Program for Population Genetics
Harvard School of Public Health
Boston, Massachusetts
USA

Nicholas J. Schork
Department of Epidemiology and Biostatistics,
Case Western Reserve University
Cleveland, Ohio
USA

Gillian Smith
University of Dundee Biomedical Research Centre
Imperial Cancer Research Fund Molecular Pharmacology Unit
Ninewells Hospital
Dundee
UK

Harold Snieder
Twin Research & Genetic Epidemiology Unit
St Thomas' Hospital
London
UK

Tim D. Spector
Twin Research & Genetic Epidemiology Unit
St Thomas' Hospital
London
UK

David Strachan
Department of Public Health Sciences
St. George's Hospital Medical School
London
UK

Pia K. Verkasalo
Department of Mental Health and Alcohol Research
National Public Health Institute
Helsinki
Finland

Anyun Wang
Anhui MEIZHONG Institute for Biomedical Sciences and
Environmental Health
Anhui
China

Binyan Wang
Program for Population Genetics
Harvard School of Public Health
Boston, Massachusetts
USA

Christopher J. Williams
Division of Statistics
University of Idaho
Moscow
Idaho
USA

C. Roland Wolf
University of Dundee Biomedical Research Centre
Imperial Cancer Research Fund Molecular Pharmacology Unit
Ninewells Hospital
Dundee
UK

Xiping Xu
Program for Population Genetics
Harvard University School of Public Health
Boston, Massachusetts
USA

PREFACE

The confluence of the fast-flowing currents of genetics and epidemiology has generated waves that have spread across both disciplines. Some ply these treacherous waters without trepidation; however, many are daunted by the seemingly unfathomable depths and by the sheer distance of the far horizon. The idea behind the First International Workshop on the Genetic Epidemiology of Complex Traits using Twins and Sib-pairs at St Catharine's College, Cambridge in 1998, was to help fellow voyagers embark but be aware of the dangers. By bringing together a disparate crew of epidemiologists, clinicians, statisticians and geneticists for two intense days in the austere but historic atmosphere of the Mill Lane lecture theatres we hoped to chart a common course.

This book has sprung from the meeting. In it, the authors have expanded on their original lectures and we have kindly received contributions from a number that were unable to attend. It has been our aim to make the subject comprehensible to all disciplines. It was not our aim to be comprehensive; but we have been keen to focus on areas of development and debate. As a result, the book contains some opposing views and perspectives.

Broadly, the chapters cover three themes. The first considers epidemiological issues concerned with collecting data on twin and sib-pairs. The chapters review the impact of the design of these studies on the interpretation of their results, ways in which to accommodate the effects of bias, the assumptions behind the twin model itself, and the use of twins to study environmental variation. Second, classical and modern approaches to estimate the genetic and environmental influences are discussed together with gene-environment interaction. Third, methods of assessing the contribution of specific genes through linkage and association are examined. The transmission disequilibrium test and the extension of variance components analysis to incorporate data from Quantitative Trait Loci (QTLs) are reviewed. Other subjects have also been interjected. These include an appraisal of the contribution of data from developing countries to twin and sib-pair studies and an assessment of the impact of phamacognetics on medicine and therapeutics.

In the editorial process, we have unashamedly used our own understanding (and lack of it) as our guide. We are grateful for the patience and restraint that authors have shown in answering our questions and in allowing us to change the format of their contributions to suit our common needs.

We recognise that it may be bold to name our first voyage an 'advance'. There was progress at least, and we are grateful that we did not founder. The success of the enterprise was entirely due to the skilful guidance and support from the seasoned travellers, and the enthusiasm of the participants.

T.D.S.
H.S.
A.J.MacG
July 1999

ACKNOWLEDGMENTS

We would like to thank all the people that made the first workshop possible including Pat Harris, Carol Roach and William Carter.

We are also grateful to our main meeting sponsors, Gemini Research Ltd of Cambridge, UK, with additional thanks to SKB and Zeneca PLC for their support.

T.D.S.
H.S.
A.J.MacG
July 1999

1

THE HISTORY OF TWIN AND SIBLING-PAIR STUDIES

Tim D. Spector

ABSTRACT

This chapter briefly discusses the origins of the classical twin study and what is today known as the sib pair method. Although Galton is rightly known as the father of behavioural and quantitative genetics and recognised that the study of twins could shed light on the effects of environment, it was unclear whether he suspected there were two biological types of twin. The first published classical twin studies appeared in 1924, simultaneously by a German dermatologist, Hermann Siemens and an American psychologist, Curtis Merriman. The sib pair approach dates back to Penrose in 1935, who first proposed studying sibs without parental information and who examined the linkage of blood groups with hair colour.

THE TWIN STUDY

"There are twins of the same sex so alike in body and mind that not even their own mothers can distinguish them . . . .This close resemblance necessarily gives way under the gradually accumulated influences of differences in nurture, but it often lasts until manhood"(Galton 1874).[1]

It appears that Galton viewed twins as useful subjects to study longitudinally through childhood to observe whether they grew more or less similar by adulthood, i.e nature versus nurture. In his 1875 article in Fraser's magazine (p 566) he noted:

"The life history of twins supplies what I wanted. We might begin by enquiring into twins who were closely alike in boyhood and youth, and who were educated together and learn whether they subsequently grew unlike . . . We can enquire in to the history of twins who were exceedingly unlike in childhood and, learn how far they became assimilated under the influence of their identical nurtures."(Galton 1875)[2]

In his observations of 35 twins of close similarity he noted some change and some similarity in their life histories, and of the 20 dissimilar twins found, 'not a single case' where they had become more similar.

Whether Galton believed that distinct types of twins existed is not clear, although he makes no clear connection with his cases of similar and dissimilar twins making it unlikely that he saw the full potential of twins.[3]

In the years that followed Galton, it appears there was considerable confusion over the biology of twinning that hampered observational research. Price,[4] in a review of much of the early twin obstetric literature, quoting the works of Meckel (1850), Schultze (1854) and Kleinwachter (1871) suggested that by the 1870's, the "one egg" origin of monochorionic twins was generally known. However the biological origin of dichorionic twins was hardly mentioned and uncertainty over the number of different types of twins persisted. This may have been because of the confusion caused by some MZ twins being dichorionic, as well as the problems of correctly identifying 'similar and non-similar' twins. Thorndike (1905)[5] was an educational psychologist who studied patterns of learning in young twins in the US and was important in propagating the one type of twin hypothesis. He found that twins even after lengthy training, did not grow more similar for certain tasks, suggesting to him that heredity was important. He also studied the skewed unimodal distribution of resemblance between all types of twins and concluded:

"The evidence in the case of the thirty-nine pairs of twins for whom we have extended physical measurements gives no reason for the hypothesis of two such distinct groups of twins." This view was later reinforced by a re-analysis of the same data by the famous RA Fisher (1919).[6]

However the prevailing opinion of the time is unclear as, Weissman in 1904[7] stated:

"It is well known that there are two kinds of twin . . . We have now every reason to believe that twins of the former kind [non-identical] are derived from two different ovum, and that those of the latter kind [identical] arise from a single ovum, which after fertilisation, has divided into two ova."

Data on the ratios of same sex males, same sex females and mixed sex twins collected during and after the first world war, (approximately 1: 1: 1) also supported the two types of twin argument, as stated by a number of authors, Danforth in 1916,[8] Newman 1917,[9] Davenport 1920,[10] and Sharp in 1921,[11] as quoted by Merriman (1924).[12]

In the same year 1924, two independent published reports appeared, one from Germany and the other from the USA outlining for the first time the principle of the existence of two types of twins that could be compared and the basis of the classical twin study. Hermann Siemens, a German dermatologist published a book (in German) on twins (*Die Zwillingspathologie 1924*)[13] in which he stated (p 21) . . .

"With the help of twin pathology we found a possible way to judge hereditary factors on the features under investigation . . . .The assessment is based on the comparison of the findings in identical and non-identical twins."

He also suggested why his work was novel.

" (p.9) The reason why this promising area of genetic research has until now never been systematically explored, may partly be due to the fact that the diagnosis of one egg twins is generally regarded as a very difficult problem."

Siemens examined the similarities of skin naevi and found a correlation of 0.4 (±0.13) in identical and 0.20 (±0.19) in non-identicals. He concluded (p 29):

> "These results are exactly as one might expect in a genetically determined naevus disposition; one may assume that there are hereditary differences in the development of the naevus."

Simultaneously, an American educational psychologist, Curtis Merriman (1924)[12] reported a study in Psychological Monographs in which he found the similarity of intelligence tests in a subgroup of identical twins (correlations ranging from 88–98%) to be 'materially' higher than the similarity of the total twin population. Although he did not explicitly identify non-identical twins for this study, he had in his introduction (p.26) listed the theory (without explaining its origins) as being: "

1. There are two distinct types of twins, fraternal and duplicate.
2. The fraternal, being of the two-egg origin, should show no greater resemblance than ordinary siblings, since each individual of the pair develops from a wholly independent arrangement of the factors for heredity in the germ cells.
3. The duplicate, being of the one-egg origin, should show a very much higher degree of resemblance than the fraternal because each member of the pair develops from substantially the same arrangement of the factors for heredity in the germ cells."

The classical twin study, as we know it, was born. It was left to others however to follow up this work and a number of classical twin papers appeared in the late twenties. The hundreds of papers that have followed since, including the full range of human characteristics and diseases are a testament to the wide applicability of the method [14] (Figure 1.1)

THE SIBLING PAIR METHOD

For the limited context of this chapter the sib pair method refers to the use of pairs of siblings without parental or other family information to determine linkage of a trait or disease to a genetic marker. The statistical basis is an assessment of the excess of allele sharing between siblings with similar phenotypes. The affected sib pair method is the most powerful statistically and the most commonly known, in the situation where both sibs have the discrete disease or trait, but the method also uses discordant sibs (only one affected) or where sibs both have a measurable continuous trait.

The statistical basis of linkage can be traced back to work at the beginning of the century as reviewed in detail by Edwards.[15] Work on non-independent segregation in the sweet pea by Bateson et al (1905)[16] led to the theory that this was due to loci close together on the same chromosome (Morgan 1911).[17] The first statistical estimator of linkage probably emanating from Engeldow & Yule's least squares method (1914),[18] followed by the better known work of Haldane (1919).[19]

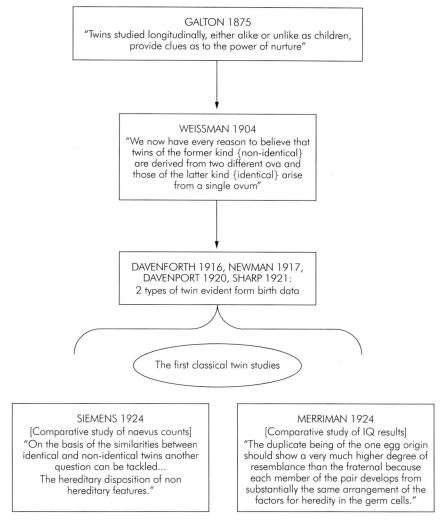

Figure 1.1 – The origins of the classical twin study

Penrose working in Colchester, UK as part of Fisher's Blood-Group Serum Unit in 1935[20] was the first to suggest using affected and unaffected sib pairs for genetic analysis without parental information. He states the benefits as being the practicalities and ease of collecting large numbers of sibships from schools, hospitals and institutions and:

> "by confining the investigation to one generation, errors due to the variation in characters with age are diminished.".

His analysis was based on the principle that:

"when pairs of sibs are taken at random from a series of families, certain types of sibling pairs will be more frequent if there is linkage than if there is free assortment of the characters studied".

To test his method he examined 60 pairs of siblings to test whether they were alike or unlike for the genetic markers, red hair, blue eyes and blood Agglutinogen A or B. He postulated that if:

"linkage exists, the number of cases where the sibs are both alike or both unlike will be relatively increased. To ascertain whether, for a given set of data, a deviation from proportionality in the classes is significant, the test described by Fisher (The exact treatment of 2x2 tables, 1934) may be conveniently used."

In his study he found one clear example of linkage between red hair and Agglutinogen A, whereby 40 pairs were alike for both, 3 pairs unlike for both and 17 like for red hair but unlike for Agglutinogen A. He estimated the likelihood of this being due to chance as 30 to 1. He suggested that this method will only be useful when the two traits or markers are common and where at least 100 pairs are available even under the most favourable of circumstances (Penrose 1935).[20]

Although work on the rhesus blood group system (shown by Fisher in 1943 to be composed of closely linked genes in linkage disequilibrium) continued, these studies in the 50s and 60s were predominantly case-control or association studies and in the search for rare Mendelian disease genes, the use of sib pair collections was temporarily forgotten.[21]

Haseman and Elston in 1972[22] proposed a theoretical basis for looking at linkage between a quantitative trait and genetic markers using a method based on regressing the difference between sibling pairs and the degree of allele sharing. This landmark paper was later recognised to be the basis for most modern linkage studies of complex traits and many statistical linkage software packages.

The first modern study to look specifically at degrees of segregation of a known genetic marker within sib pairs with a discrete trait or disease was probably published in a short letter in French in 1974 by Hors and Dausset[23]. They examined the sharing of HLA-A antigens within 23 affected sib pairs with a variety of cancers and found a non-significant 2–3 fold increase in HLA sharing compared to that expected, but called for more patients. The next year, Bobrow et al (1975)[24] in a pilot study of 9 families containing two affected sib pairs found no evidence of linkage between HLA and spina bifida. At the same time Cudworth and Woodrow (1975)[25] found evidence of significant linkage between HLA-A8 and W15 with Type I Diabetes using 17 families and 15 affected sib pairs. Ten of the 15 pairs had identical HLA haplotypes, which was highly significant. However this study used parental information to determine haplotypes and is not therefore a true affected sib pair study in the modern sense. Nevertheless these were the first studies to use the advantage of the affected sib pair approach which avoided the problems of case definition in unaffected relatives and also allowed the analysis to be performed independent of the underlying genetic model.

The underlying statistical theory of perturbations in IBD distribution at marker loci followed the clinical studies, initially by Day & Simons (1976)[26] and Suarez et al (1978).[27] Since then the affected sib pair study has blossomed and recently extended to include combined tests of linkage and association (sibling transmission disequilibrium test S-TDT) which allow the use of sibling information without the need for parental details in both discrete (Spielman and Ewens 1998)[28] and quantitative traits (Fulker et al 1999)[29] (Figure 1.2)

In summary, from an overview of the published literature it appears that the sib pair design owes its origins primarily to the work of Penrose in 1935 and the classical twin pair design to Merriman and Siemens in 1924.

ACKNOWLEDGEMENTS

I would like to thank Harold Snieder for his helpful comments and German translations, Christel Barnetson for help with the figures and numerous colleagues for references.

Date	Authors	Comments
1905	Bateson	"Non-independent segregation of loci"
1911	Morgan	"Loci must be close together"
1914	Engledow & Yule	"Least square estimators at linkage"
1919	Haldane	"Extension of linkage methods"
1935	Penrose	"First use of sib pairs without parents to determine linkage between red hair and blood group A"
1972	Haseman & Elston	"Theoretical use of sibs for continuous traits"
1974	Hors & Dausset	"Affected sib pair study of cancer & HLA-A"
1975	Bobrow et al Cudworth & Woodrow	"Affected sib pair study of spina bifida & HLA-A" "Affected sib pair study of diabetes & HLA-A"
1976	Day & Smith	"Statistical justification of IBD distribution method"
1978	Suarez et al	"Statistical justification of IBD distribution method"
1998	Spielman & Ewens	"Sib-sib TDT for discrete traits"
1999	Fulker et al	"Sib-sib TDT for continuous traits"

Figure 1.2 – Landmarks in the history of the sib pair study

REFERENCES

1 Galton F. English men of science: Their nature and nurture. Clarke, Doble & Brendon 1874, London.

2 Galton F. The history of twins, as criterion of the relative powers of nature and nurture. Fraser's magazine 1875, Nov.: 566–576

3 Rende RD, Plomin R, Vandenberg SG. Who discovered the twin method? Behaviour Genetics, 1990;20: 277–285

4 Price B. Primary biases in twin studies: a review of the prenatal and natal difference-producing factors in monozygotic pairs. Am J Human Genetics 1950;2: 293–352

5 Thorndike EL. Measurement of twins. Archives of Philosophy, Psychology and Scientific Methods, 1905; 1: September.

6 Fisher RA. The genesis of twins. Genetics 1919;4: 489–499

7 Weismann A. The evolutionary theory 1904;2: 44–45 (quoted by Merriman)

8 Danforth CH. Is twinning hereditary? Journal of Heredity 1916;7: 195

9 Newman HH. The biology of twins. University of Chicago Press 1917, pp 1–185

10 Davenport CB et al. Twins. Journal of Heredity. 1919: december (quoted by Merriman)

11 Sharp. Introduction to Cytology, 1921: p357 (quoted by Merriman)

12 Merriman C. The intellectual resemblance of twins. Psychological Monographs 1924;33: 1–58

13 Siemens HW. Zwillingspathologie: Ihre Bedeutung; ihre Methodik, ihre bisherigen Ergebnisse. 1924 Berlin. Springer Verlag

14 Neale MC, Cardon LR. Methodology for genetic studies of twins and families. Nato ASI series vol 67, Kluwer Academic Press, Dordrect, 1989 pp 26–33

15 Edwards AWF. The early history of the statistical estimation of linkage. Ann Hum. Genetics 1996;60: 237–249

16 Bateson W, Saunders ER, Punnett RC. (1905) Reports to the evolution committee of the Royal Society. II Experimental studies on the physiology of heredity. London

17 Morgan TH. Random segregation versus coupling in Mendelian Inheritance. Science 1911;34: 384

18 Engledow FL, Yule GU. The determination of the best value of the coupling ratio from a given set of data. Proc Camb Phil. Soc. 1914;17: 436–440.

19 Haldane JBS. The combination of linkage values, and the calculation of distances between the loci of linked factors. Journal of Genetics 1919;8: 299–309

20 Penrose LS. The detection of autosomal linkage in data which consist of pairs of brothers and sisters of unspecified parentage. Ann Eugenics 1935;3: 133–138

21 Tomlinson IPM, Bodmer WF. The HLA system and the analysis of multifactorial genetic disease. Trends in Genetics 1995;11: 493–498

22 Haseman JK, Elston RC. The investigation of linkage between a quantitative trait and a marker locus. Behaviour Genetics 1972;2: 3–19

23 Hors J, Dausset J. Maladies malignes identiques au sein d'une meme famille. La Nouvelle Presse medicale 1974;3: 1237

24 Bobrow M, Bodmer JG, Bodmer WF, McDevitt HO, Lorber J, Swift P. The search for the human equivalent of the mouse T locus- negative results from a study of HL-A types in Spina Bifida. Tissue Antigens 1975;5: 234–237

25 Cudworth AG, Woodrow JC. Evidence for HL-A-linked genes in Juvenile Diabetes Mellitus. British Medical Journal 1975;3: 133–135

26 Day NE, Simons MJ. Disease susceptibility genes – Their identification by multiple case family studies. Tissue Antigens 1976;8: 109–119

27 Suarez BK, Rile J, Reich T. The generalised sib-pair IBD distribution: its use in the detection of linkage. Ann Human Genetics London 1978;42: 87–94

28 Spielman RS, Ewens WJ. A Sibship test for linkage in the presence of association: The sib transmission / disequilibrium test. 1998;62: 450–458

29 Fulker DW, Cherney SS, Sham PC, Hewitt JK. Combined Linkage and Association sib-pair analysis for quantitative traits. Am J Hum Genetics 1999 64: 259–267.

2

INFLUENCE OF DESIGN ON THE OUTCOME OF TWIN STUDIES

David P. Strachan

ABSTRACT

The principal methods of identifying twins for research purposes are through birth records; selected populations such as military recruits; media appeals; or among patients presenting with clinical disease. Twin concordance studies and co-twin comparisons published over the last 30 years have been based mainly on disease ascertainment by questionnaire, interview or objective testing among all members of a twin register (34 studies based on birth cohorts, 16 on selected populations and 15 on volunteer series); linkage of routine data to a twin register (50 studies); and recruitment of twins with a disease of interest from clinical case-series (47 studies). Various other designs accounted for 33 studies.

The disease definitions applied in probands and co-twins are influential in determining absolute concordance rates, and may also be important in comparisons of monozygotic and dizygotic concordance rates. The assumption that probandwise concordance rates are unbiased provided there is complete ascertainment of disease in co-twins is not wholly valid in many twin studies, due to non-independent ascertainment (information) bias; concordance-dependent (selection) bias; and differential enrolment (selection) bias.

Few studies have compared results from different methods of twin iden- tification or disease ascertainment, either within or between studies. More comparisons of this type are needed if the results from less compre- hensive methods of identification or ascertainment are to be interpreted with greater confidence. However, few twin studies are entirely free from bias and the consequences of each design should be considered for each study.

This chapter outlines the scope of twin studies and reviews the relative advantages and disadvantages of the principal twin study designs. The main sources of bias arising in studies of twin concordance are introduced and some of these will be elaborated in later chapters. The emphasis here is on studies of the presence or absence of clinical disease, whereas later chapters will concentrate to a greater extent on traits which can be measured on a continuous scale.

Why study twins?

It has long been recognised that twins offer unusual opportunities for biomedical and psychosocial research.[1,2] The comparisons which may be informative include:

i) Disease incidence, prevalence or outcome in twins compared to singletons.

ii) Disease concordance rates, or correlations of continuous traits in monozygotic compared to dizygotic twins.

iii) Observed concordance rates in either type of twin compared to the degree of concordance expected from the disease incidence or prevalence among twins.

iv) Co-twin cohort studies or controlled trials comparing disease rates in twins who are discordant for developmental, lifestyle, environment or medical care factors.

v) Co-twin case-control studies comparing levels of exposure to potential causes of disease in twins (especially monozygotic pairs) who are discordant for disease.

vi) Co-twin natural history studies of early markers or biological mechanisms of disease in the unaffected member of disease-discordant monozygotic pairs.

vii) Genetic linkage and association studies using dizygotic pairs as a special application of the sib-pair design.

Twin studies may arise in a variety of ways. Twins affected by a disease of interest may be identified from a clinical case-series or a more or less comprehensive disease register. Alternatively, diseases and other characteristics of interest may be assessed among sets of twins responding to media appeals or among twin pairs ascertained in a more or less systematic manner from birth records or other population registers.

Identification of twins

The principal methods of identifying twins for research purposes are:

i) A complete population register, usually based on birth records. This approach underlies the nationwide twin registries in Scandinavian countries,[3-7] and is also the basis of some local twin registers.

ii) A selected population among whom twins have been systematically identified. For example, US twin registers have been based on military recruits[8,9] and on patients registered with the Kaiser-Permanente health maintenance organisation.[10]

iii) A media appeal for twin "volunteers". This approach underlies the Australian nationwide twin registry[11] and several local twin series.

iv) Twins presenting to clinicians with a disease of interest.

v) A media appeal for twins with a particular disease or characteristic.

vi) Special groups, such as twins reared apart.[12,13]

Ascertainment of disease

The principal methods of ascertaining disease in twin studies are as follows:

i)   Assessment of all twins by questionnaire, interview or objective tests.
ii) Record linkage of routinely ascertained events such as death, disease regis-
     tration or hospital admission.
iii) Patients with specific diseases of interest presenting to clinicians.
iv) Mutual support associations for patients with specific diseases of interest.
v)   A media appeal for twins with a particular disease.

Although this chapter will concentrate on clinical diseases, the same principles
apply to developmental, psychological, behavioural, educational and sociological
outcomes.

Classification of twin study designs

Table 2.1 summarises the twin concordance and co-twin studies published during
the past 30 years, in terms of the methods used to identify twins and to ascertain
disease. These studies were identified by a Medline search for the years 1966 to
mid-1997, using textword strings [*twin* and (*concord* or *discord*)]. Studies of obstetric
complications, congenital abnormalities, review articles and methodological papers
are excluded from the table. The search strategy was not designed to identify trait
correlation studies but should give a reasonably accurate reflection of the methodology applied in disease concordance and co-twin studies. The most commonly
used study designs involve:

i)   Disease ascertainment by questionnaire, interview or objective testing among
     all members of a twin register. The largest group were based on a compre-
     hensive population twin register: 24 concordance studies and 30 co-twin

*Table 2.1 – Number of concordance and co-twin studies of clinical disease published
1966 to 1997, by source of twins and method of ascertainment of disease.*

Source of twins	Questions or tests		Ascertainment of disease: Record linkage		Clinical cases		Other methods		
Study type:	Con	Cot	Con	Cot	Con	Cot	Con	Cot	Total
Population	24	30	26	16	1	1	-	–	98
Army or HMO	10	6	3	5	1	–	–	–	25
Media appeal	10	5	-	-	-	-	4	1	20
Clinic patients	1	-	1	1	18	29	6	11	67
Reared apart	5	-	-	-	-	-	-	–	5
Total	50	41	30	22	20	30	10	12	215

Con = concordance study   Cot = Co-twin study

studies. Selected population registers were the basis of 10 concordance studies and 6 co-twin studies, and volunteer series generated 10 concordance studies and 5 co-twin studies.

ii) Linkage of routine data to a twin register – 29 concordance studies and 21 co-twin studies.

iii) Recruitment of twins with a disease of interest from clinical case-series – 18 concordance studies and 29 co-twin studies.

iv) Various other designs accounted for 19 concordance studies and 14 co-twin studies.

MEASUREMENT OF TWIN CONCORDANCE

Identification of twin pairs which are concordant (ie both members are affected by a disease or characteristic of interest) has important implications both for concordance studies and for selection of discordant pairs for co-twin studies. Three issues are highly influential in determining absolute concordance rates, and may also be important in comparisons of monozygotic and dizygotic concordance rates:

i) The disease definition used for ascertainment of probands.

ii) The criteria applied to determine presence or absence of disease in co-twins of affected probands.

iii) The time period over which these criteria are applied.

Ascertainment of probands

In studies which rely on routine data linkage, or where the source of twins is a clinical case-series, then probands are by definition cases of diagnosed illness. However, when a questionnaire or objective testing is applied to a registry sample, there is an opportunity to choose between a sensitive, inclusive case definition (leading to higher prevalence but less severe disease), or a more specific, exclusive definition (leading to lower prevalence and usually more severe disease).

Several studies have reported twin concordance results for specific diseases with different case-definitions, for example Kendler *et al.*[14] used three definitions of alcoholism: an exclusive definition (dependence-tolerance), intermediate (alcoholism with or without dependence), and an inclusive definition (any problem drinking). Monozygotic concordance rates varied from 26% with the exclusive definition to 47% for problem drinking; the dizygotic concordance rates were, respectively, 12% and 32%. Heritability estimates were also dependent on case-definition, varying from 50% for dependence-tolerance to 61% for problem drinking.

Assessment of co-twins

Co-twins of affected probands can usually be assessed more thoroughly for the presence or absence of disease than can the entire population of twins. This allows greater flexibility in disease definition and for calculation of concordance rates more inclusive criteria are preferred. To a first approximation,[15] valid concordance rates can be derived even if ascertainment of probands is incomplete, provided that disease detection in co-twins is complete (this assumption is not always justified, as discussed below). However, a wide range of minor symptoms have been reported for co-twins of probands with alcoholism[16] or schizophrenia,[17] and this generates uncertainty in determining which co-twins are affected by the disease.

An inclusive disease definition is also preferred in selection of disease-discordant pairs for co-twin case-control studies.[18] In this context, it may be appropriate to exclude also co-twins with markers of early or subclinical disease, such as glucose intolerance in studies of diabetes.[19] The advantages of selecting extremely discordant pairs in this manner are that diagnostic misclassification is minimised, improving the power to detect associations with causal factors. The disadvantages are that it reduces the number of pairs eligible for the study and limits the generalisability of the findings. Furthermore, exclusion of co-twins with early disease may lead us to overlook causal factors which operate late in a multi-stage process, or which modulate the progression from asymptomatic to symptomatic disease.

Longitudinal studies

The absolute level of concordance can be strongly influenced by the time interval over which cases of disease in the co-twin are ascertained. The incidence of clinical disease and subclinical abnormalities is the focus of co-twin natural history studies. For instance, in a study of 30 identical co-twins of children with insulin-dependent diabetes,[20] 7 were concordant for the disease at the time the proband was ascertained, but a further 7 developed clinical diabetes 3–36 years later. Among the 16 twins who remained non-diabetic, 4 developed abnormally low insulin release after 15–26 years, and another 4 developed pancreatic islet-cell antibodies during follow-up. Other longitudinal studies of childhood diabetes show a similar pattern of delayed onset of clinical or subclinical disease in a proportion of co-twins.[21,22]

In general, concordance rates will be biased upwards if only the co-twins of affected probands are followed over time, because new cases will emerge in previously unaffected twin pairs during observation period, leading to an increase in the number of discordant pairs, who should also be included in any calculation of concordance rates.[23] For this reason, disease status should either be ascertained in all twins over a specified period of follow-up, or among co-twins only at the time of disease ascertainment in the proband.

Potential biases in twin concordance studies

The assumption that probandwise (casewise) concordance rates (as opposed to pairwise concordance rates) are unbiased provided there is complete ascertainment of disease in co-twins[24] is not wholly valid in many twin studies (See chapter 4 Box 1. Three influences have been considered in detail from a theoretical viewpoint:

i) Non-independent ascertainment (information) bias. The above assumption purportedly allows for the possibility that co-twins of affected probands are more likely to be ascertained than co-twins of unaffected individuals. However, Kendler and Eaves[24] show that the probandwise concordance can be biased if there is a large difference in proband and co-twin ascertainment rates, particularly if the true concordance rate is high. Thus, clinical case-series (in which proband ascertainment is usually incomplete and unquantified) may give misleadingly high concordance rates, compared to studies where ascertainment is comprehensive (as in questionnaire surveys of twin register samples). Record linkage studies, which rely on routinely recorded health events, are also prone to problems due to non-independent ascertainment.

ii) Concordance-dependent (selection) bias. Concordance will be biased upwards if disease-concordant twin pairs are more likely to be ascertained than discordant pairs.[25] The influence on estimates of heritability will be particularly strong if concordant-dependent selection occurs to a greater degree in monozygotic pairs than dizygotic pairs. The extent to which this occurs is rarely measurable, but on theoretical grounds it is likely to be more common in clinical case-series, due selective referral of "unusual" cases. The more comprehensively the disease is ascertained, the less opportunity there is for concordant-dependent selection. The impact of this form of bias is greater when the true concordance is low.[24]

iii) Differential enrolment (selection) bias. The prevalence of disease among the twin sample as a whole may be biased upwards or downwards, if twins with disease are more or less likely to be registered. This in turn affects the concordance rates in both monozygotic and dizygotic twins and may influence heritability estimates.[26] This phenomenon is of particular concern in registers based on twin volunteers[27] or groups subjected to health-related selection, such as military recruits.[26] Although the direction of the bias is predictable, its magnitude may be difficult to estimate without information on the prevalence of disease among twins not included on the register.

ADVANTAGES AND DISADVANTAGES OF THE PRINCIPAL STUDY DESIGNS

Bearing in mind these possible influences on measures of twin concordance, Table 2.2 summarises the advantages and disadvantages of population registers, volunteer series and diseased twins, and the relative strengths and weaknesses of the most widely used methods of disease ascertainment.

Has design been shown to influence concordance?

Few studies can be identified which have compared different methods of twin identification or disease ascertainment in terms of their effect on concordance rates. This is presumably because a single study design is almost invariably chosen at the outset due to practical constraints, and the desire to standardise procedures throughout the study. Comparisons may be made within studies or between studies.

An example of such a comparison within the same study is provided by Silman *et al.*[28] in a study of rheumatoid arthritis. Affected twin pairs were recruited by multi-media advertisement and through collaborating rheumatology clinics. The concordance rates for monozygotic twins were similar in the two series (16% and 13%, respectively). The dizygotic concordance rates (3% and 6%, respectively) were as similar as might be expected with the relatively small sample size involved.

A second example assessed whether concordance rates were affected by restricting comparisons to twin pairs who remained resident in the same geographical area. In the Finnish twin registry,[29] concordance rates for psychosis differed markedly between pairs living in the same area (monozygotic concordance: 27%, dizygotic concordance: 6%) and pairs living in different areas (concordance rates 0% and 1%, respectively). However, for a contrasting condition, hypertension, the concordance rates were much more similar in the two groups (54% v 47% for monozygotic twins, 32% v 26% for dizygotic twins).

Table 2.2 – Relative advantages and disadvantages of different twin study designs.

Advantages	Disadvantages
Clinical case-series	
No requirement for twin register	No estimate of disease prevalence
Efficient, particularly for rare diseases	Selective ascertainment of concordant pairs
Comprehensive for twin status (& zygosity)	Arbitrary & inflexible case definition
Volunteer twin series	
No requirement for twin register	Bias towards concordant pairs
Higher response to surveys or tests	Unrepresentative prevalence figures
Flexibility in case definition	Zygosity may be incompletely confirmed
Population-based twin register	
More representative prevalence figures	Often difficult to set up and maintain
No inherent bias towards concordant pairs	Incomplete response may bias prevalence
Flexibility in case definition	Zygosity may be incompletely confirmed
Record linkage to routine data	
Highly efficient if available	Ascertainment may be incomplete
Usually representative	Not immune to biases in concordance
Comparison of twins v singletons	Inflexible case definition
Questionnaires or tests	
Ascertainment systematic (may be incomplete)	Non-response may bias prevalence
Flexible case definition (& objectivity if tests)	Inefficient (especially if tests used)
Less prone to concordance-related biases	

A third within-study comparison is reported by Kendler *et al.*[30] who compared questionnaire and record linkage as methods of ascertainment for affective disorder within the Swedish twin registry. The monozygotic concordance rate for a broad definition of affective illness was somewhat higher for probands ascertained through hospital admission records (77%) compared to community ascertainment (62%). The dizygotic concordance rates were similar (33% and 37%, respectively).

Another approach is to compare the results of different studies which have adopted different designs. Walker *et al.*[31] report a meta-analysis of 11 twin concordance studies of schizophrenia and 10 studies of affective disorder. Three features of study design were examined as independent predictors of monozygotic and dizygotic concordance rates: objective zygosity testing, source of twins (clinical series *versus* registry sample), and more recent publication. No consistent pattern emerged: all three factors influenced dizygotic concordance rates for affective disorder, but none predicted dizygotic concordance for schizophrenia. Clinical case-series had higher monozygotic concordance rates for schizophrenia but not for affective disorder, whereas studies using objective zygosity testing had higher monozygotic concordance rates for affective disorder, but not for schizophrenia.

CONCLUSIONS

Many theoretical considerations suggest that the design of twin studies should influence the outcome, particularly the measurement of concordance rates and the estimation of heritability. However, there are remarkably few examples in the published literature where these predictions have been tested. More comparisons of this type are needed if the results from less comprehensive methods of twin identification or disease ascertainment are to be interpreted with greater confidence. However, few twin studies are entirely free from bias and the consequences of each design should be considered for each study in its own context.

REFERENCES

1 Hrubec Z, Robinette CD. The study of human twins in medical research. *New England Journal of Medicine* 1984;**310**: 435–441.

2 Anon. The use of twins in epidemiological studies. Report of the WHO meeting of investigators on methodology of twin studies. *WHO Chronicle* 1966;**20**: 121–128.

3 Medlund P, Cederlöf R, Floderus-Myrhed B, Friberg L, Sörensen S. A new Swedish twin registry. *Acta Medica Scandinavica* 1976;**Suppl 600**: 1–104.

4 Kringlen E. Norwegian twin registers. (in) Nance WE (ed) *Twin Research. Proceedings of the second international congress of twin studies. Part B. Biology and epidemiology.* New York: Alan R Liss, 1978: 189–195.

5 Kaprio J, Sarna S, Koskenvuo M, Rantasalo I. The Finnish twin registry: formulation and compilation, questionnaire study, zygosity determination, procedures and research program. *Progress in Clinical & Biological Research* 1978;**24B**: 179–184.

6 Hauge M, Harvald B, Fischer M, Gotlieb-Jensen K, Juel-Nielsen N, Raebild I, Shapiro R, Videbech T. The Danish twin register. *Acta Geneticae Medicae et Gemellologiae* 1968;**2**: 315–331.

7 Kyvik KO, Green A, Beck-Nielsen H. The new Danish twin register: establishment and analysis of twinning rates. *International Journal of Epidemiology* 1995;**24**: 589–596.

8 Jablon S, Neel JV, Gershowitz H, Atkinson GF. The NAS-NRC twin panel: methods of construction of the panel, zygosity diagnosis, and proposed use. *American Journal of Human Genetics* 1967;**19**: 133–161.

9 Tsuang MT, Lyons MJ, Eisen SA, Goldberg J, True W, Lin N, Meyer JM, Toomey R, Faraone SV, Eaves L. Genetic influences on DSM-III-R drug abuse and dependence: a study of 3,372 twin pairs. *American Journal of Medical Genetics* 1996;**67**: 473–477.

10 Friedman GD. Twin studies of disease heritability based on medical records: application to acne vulgaris. *Acta Geneticae Medicae et Gemellologiae* 1984;**33**: 487–495.

11 Duffy DL, Martin NG, Battistutta D, Hopper JL, Mathews JD. Genetics of asthma and hay fever in Australian twins. *American Review of Respiratory Disease* 1990;**142**: 1351–1358.

12 Gatz M, Pedersen NL, Berg S, Johansson B, Johansson K, Mortimer JA, Posner SF, Viitanen M, Winblad B, Ahlbom A. Heritability for Alzheimer's disease: the study of dementia in Swedish twins. *Journals of Gerontology* 1997;**Series A, Biological**: M117-M125.

13 Hankins D, Drage C, Zamel N, Kronenberg R. Pulmonary function in identical twins raised apart. *American Review of Respiratory Disease* 1982;**125**: 119–121.

14 Kendler KS, Heath AC, Neale MC, Kessler RC, Eaves LJ. A population-based twin study of alcoholism in women. *Journal of the American Medical Association* 1992;**268**: 1877–1882.

15 Allen G, Hrubec Z. Twin concordance. A more general model. *Acta Geneticae Medicae et Gemellologiae* 1979;**28**: 3–13.

16 Pickens RW, Svikis DS, McGue M, Lykken DT, Heston LL, Clayton PJ. Heterogeneity in the inheritance of alcoholism. A study of male and female twins. *Archives of General Psychiatry* 1991;**48**: 19–28.

17 Onstad S, Skre I, Torgersen S, Kringlen E. Twin concordance for DSM-III-R schizophrenia. *Acta Psychiatrica Scandinavica* 1991;**83**: 395–401.

18 de Faire U, Lundman T. Death-discordant twins-a new method to evaluate genetic factors in chronic diseases. *Acta Geneticae Medicae et Gemellologiae* 1976;**25**: 114–116.

19 Poulsen P, Vaag AA, Kyvik KO, Moller Jensen D, Beck-Nielsen H. Low birth weight is associated with NIDDM in discordant monozygotic and dizygotic twin pairs. *Diabetologia* 1997;**40**: 439–446.

20 Verge CF, Gianani R, Yu L, Pietropaolo M, Smith T, Jackson RA, Soeldner JS, Eisenbarth GS. Late progression to diabetes and evidence for chronic beta-cell autoimmunity in identical twins of patients with type I diabetes. *Diabetes* 1995;**44**: 1176–1179.

21 Olmos P, O'Hern R, Heaton DA, Millward BA, Risley D, Pyke DA, Leslie RD. The significance of the concordance rate for type 1 (insulin-dependent) diabetes in identical twins. *Diabetologia* 1988;**31**: 747–750.

22 Kyvik KO, Green A, Beck-Nielsen H. Concordance rates of insulin dependent diabetes mellitus: a population based study of young Danish twins. *British Medical Journal* 1995;**311**: 913–917.

23 Newman B, Selby JV, King MC, Slemenda C, Fabsitz R, Friedman GD. Concordance for type 2 (non-insulin-dependent) diabetes mellitus in male twins. *Diabetologia* 1987;**30**: 763–768.

24 Kendler KS, Eaves LJ. The estimation of probandwise concordance in twins: the effect of unequal ascertainment. *Acta Geneticae Medicae et Gemellologiae* 1989;**38**: 253–270.

25 Holm NV. A note on ascertainment probability in the Allen/Hrubec twin model. *Acta Geneticae Medicae et Gemellologiae* 1983;**32**: 37–47.

26 Kendler KS, Holm NV. Differential enrollment in twin registries: its effect on prevalence and concordance rates and estimates of genetic parameters. *Acta Geneticae Medicae et Gemellologiae* 1985;**34**: 125–140.

27 Lykken DT, Tellegren A, DeRubies R. Volunteer bias in twin research: the rule of two-thirds. *Social Biology* 1978;**25**: 1–9.

28 Silman AJ, MacGregor AJ, Thomson W, Holligan S, Carthy D, Farhan A, Ollier WE. Twin concordance rates for rheumatoid arthritis: results from a nationwide study. *British Journal of Rheumatology* 1993;**32**: 903–907.

29 Romanov K, Koskenvuo M, Kaprio J, Sarna S, Heikkila K. Selection bias in disease-related twin studies. Data on 11,154 adult Finnish twin pairs from a nationwide panel. *Acta Geneticae Medicae et Gemellologiae* 1990;**39**: 441–446.

30 Kendler KS, Pedersen N, Johnson L, Neale MC, Mathe AA. A pilot Swedish twin study of affective illness, including hospital- and population-ascertained subsamples. *Archives of General Psychiatry* 1993;**50**: 699–700.

31 Walker E, Downey G, Caspi A. Twin studies of psychopathology: why do the concordance rates vary? *Schizophrenia Research* 1991;**5**: 211–221.

3

SAMPLE SELECTION AND OUTCOME DEFINITION IN TWIN STUDIES: EXPERIENCES FROM THE FINNISH TWIN COHORT STUDY

Jaakko Kaprio
Pia K. Verkasalo
Markku Koskenvuo

ABSTRACT

The purpose of this chapter is to review some of the issues related to study design that investigators should consider. The chapter is not exhaustive but draws on our experiences in conducting the Finnish twin cohort studies.

The study hypothesis should guide sample selection and outcome definition, so that the study has adequate statistical power and makes efficient use of the data that is collected. The study sample of twin/sib pairs should aim to be representative of the target population which does not, however need to be a random sample of the general population. The criteria used to select the study population should be kept in mind when the results are analysed and conclusions made. An evaluation of the extent of coverage of cases and degree of misclassification of outcomes should be part of the study. Likewise, incidence data are always to be preferred. Information on twinship, zygosity, exposure and covariate data should be assessed independently of outcome status. Though each study is unique, there are many advantages in using diagnostic and operational criteria that are widely accepted and in research use. Similarly, standardised questionnaires and established laboratory procedures should always be given preference over local variants. To a large extent the procedures in twin studies for sample selection and outcome definition are the same as those used in other epidemiological research. Certain issues specific to twin studies are reviewed and illustrated in this chapter.

SCOPE OF TWIN STUDIES

While many diseases aggregate in families, few follow simple Mendelian patterns of inheritance. The resolution of whether the pattern of familiality is due to shared genes or due to factors (considered in the aggregate) shared by family members (such as, for example, common household environment, dietary habits or common beliefs and values) can be readily carried out by the classical twin study [1]. The resolution of genetic architecture has as its first goal the quantitative estimation of the overall role of genetic factors contributing to disease or trait variability. This quantitative estimation comes about through the comparison of the similarity of monozygotic (MZ) and dizygotic (DZ) twins. Identification of specific disease loci, which includes assessing the respective gene polymorphisms, allele frequencies and associated average effects, is generally the next goal. This work can also be carried out on unrelated individuals, but also on sib pairs such as either concordant or discordant twin pairs[2]. Furthermore, developments in statistical methods in recent years have permitted more sophisticated analyses through extensions of the classical twin design. Thus, data from other types of family members, such as parents, can also be used to carry out twin-family studies.[1] Methods to analyse multivariate data (either different variables, repeated measures or both) have also developed, which has increased the general requirements for data quality. Careful attention should be paid to how samples are defined, how subjects are recruited into studies, and how outcomes are defined.

Twins can also be used to test epidemiological causal hypotheses. An association between an observed exposure and the outcome of interest may be either causal or due to confounding by unknown or non-measurable factors. It may be logistically impossible or unethical to test such an association experimentally in an intervention study or randomized clinical trial. Thus, twin pairs discordant for outcome or disease are one alternative design for testing such causal hypotheses. Issues in discordant pair analyses are discussed in more detail in Chapter 5.

SAMPLE SELECTION

Twins as a minority of the target population

Sample selection starts with the definition of the target population, to which the results of the study will be generalized. From the target population, a study population is selected consisting of those who fulfil the selection criteria, such as age, gender or other characteristics. The study sample then consists of those study population subjects who are finally studied. With regard to twin and sibpair studies, it is necessary to note that twins in the target population make up only a minority of the population. Concerns that twins are not representative of the entire population have often been raised, but these should be considered case by case depending on the exposures, confounders and outcomes to be studied. Thus, e.g. the birth weights of twins are consistently lower on average than those of singletons, while we have found that smoking rates in adult twins and singletons do not differ[2]. Thus, the issue of representativeness of the twin sample with respect to all subjects, whether twins or not (see chapter 6), should be kept separate from the more universal issues of representativeness as illustrated in the Figure.

Twins versus ordinary siblings

While ordinary sibs are much more common than twin pairs, there are several advantages in studying twins rather than sibs. Twins are often easier to identify from birth records than ordinary sibs, and an existing register with twin and/or family data may therefore facilitate subject selection. Discordant twin pairs can be considered as a matched case-control series. DZ pairs are sib-pairs that are not only matched on age, but also matched on cohort effects, e.g. the number of older and younger sibs, family environment, school and cultural factors, which may be particularly important in studies of behavioural traits and psychopathology. Twin sib pairs have also a lower probability of false paternity compared to ordinary sibs. Possible disadvantages of using twins instead of ordinary sibs include the effects of possible intra-pair interaction in utero[3] and later in childhood as well as the fact that DZ twinning is a heritable trait.

Existence of concordantly affected pairs

Affected sib pairs for any disease are relatively rare, and thus there are extra costs of time and money of identifying such pairs compared to identifying series of individuals with the outcome of interest and appropriate controls. The relative rarity of concordant pairs can be seen in Table 3.1, which shows the numbers of concordant and discordant pairs affected with certain common diseases among the opposite sex twin pairs participating in the 1996–1997 survey (see boxed text for a description of the Finnish Twin Cohort data sets). Even for the most common disorder, hypertension, less than 100 concordant pairs out of more than 1500 pairs were found. The extra effort to screen for concordant twin pairs may offset the advantages of studying such pairs. Nonetheless, if the twins are already identified, the extra effort may be relatively low. We have found a higher participation of twins compared to singletons in our Finnish studies, probably because the twins are aware of their uniqueness, and it is easier to motivate participation of both twins, compared to getting two sibs to participate. Identifying concordant pairs increases the probability of finding multiplex (many affected members) families, which may be a further reason for seeking out such pairs.

Table 3.1 – Number of unaffected, discordant and concordant pairs for selected chronic diseases reported by twins from opposite-sex pairs replying to a health questionnaire in 1996–1997.

Disease	Total	Neither affected	One affected (%)	Both affected (%)
Migraine	1455	1002	408 (28.0)	45 (3.1)
Hypertension	1544	1073	386 (25.01)	85 (5.5)
Asthma	1417	1287	123 (8.6)	7 (0.5)
Obesity (BMI>30)	1544	1178	317 (20.5)	49 (3.2)

The Finnish Twin Cohorts

The older Finnish Twin Cohort consists of like-sexed pairs born prior to 1958 with both members of a pair alive in 1967. The selection procedures, determination of twinship and assessment of representativeness are described elsewhere [4]. A total of 13,888 pairs of known zygosity have been identified.

Zygosity was determined by the questionnaire method and was validated by genetic markers in a subsample[5]. Questionnaire studies have been carried out in 1975, 1981 and 1990 on the entire twin panel. Also, we have identified a cohort of younger twins born 1958–1986 [2], with two longitudinal twin-family studies in progress.

There are very few large series of adult twin pairs, which include opposite sex pairs. To identify such pairs in Finland, we applied the same selection criteria as for the older twin cohort of like-sexed pairs [4]. Permission to form the cohort was obtained from the Central Population Registry (CPR) in 1996. The computerised person file of the CPR was used to form pairs (N5017) of persons that were a) born 1938–1949, b) born in same local community, c) with the same surname at birth, d) with the same birth date, e) of opposite sex, and f) alive in 1974. Thus, these criteria yield biological twin pairs, but also pairs of persons satisfying these criteria, but who are not sibs. During June 1996 to November 1997, we have mailed a questionnaire to those pairs with both presumed co-twins alive and with a current address in Finland (N=3453 pairs). The questionnaire included items on demographic variables including twinship, personal history of selected diseases, smoking, alcohol use, weight and height, social support, and family history of selected diseases. Such information was available for 1455 pairs born at present. The data collection process is still in progress, and for pairs (N=884) in which neither twin candidate has replied so far, twinship will be confirmed from local parish records as done for the like-sexed pairs.

Generalisability of twin studies

The selection criteria that are used to define the study population will also affect the generalizability of results, as in any study. For example, the concordance of like-sexed and opposite-sexed twin pairs for migraine based on the 1990 and 1996–7 questionnaire studies illustrates this well[6] The difference in MZ and DZ concordance among both male and female like-sexed twins was very similar, suggesting that the contribution of genetic effects to inter-individual variability in susceptibility to migraine is of the same magnitude in men and women. The concordance for opposite-sex pairs was smaller than either of the like-sex DZ pair concordances, which suggests that there may be nonetheless sex-specific genetic effects.

Another example is based on the Swedish Twin Registry. Coronary heart disease mortality is a trait, where both the incidence of disease and the genetic liability are highly age-dependent. Marenberg et al[7] showed that the genetic component in

CHD mortality decreased with age, being greatest in men aged under 55 years, and least in men aged 75 years or over. Thus, study results may or may not be generalizable to sections of the population excluded by the selection criteria of one's own study. These two examples illustrate the dependence of estimates of the role of genetic and environmental factors on basic epidemiological determinants such as sex and age.

ISSUES OF OUTCOME DEFINITION

In any genetic epidemiological study, cases (defined here as subjects with the outcome of interest) should be both reliably and validly defined. Reliable measurements are highly reproducible and thus minimise error from random variation . Validity is defined as the absence of systematic error (bias), and is often divided into internal and external validity[8-10]. An aspect peculiar to genetic studies is the need to differentiate between those cases in a family (pedigree) that bring a family to the investigator's attention (the proband case) from those found in the family after the proband is identified[11]. In the study population, cases may be identified from pre-existing sources, or by conducting new data collection.

Use of existing data sources to identify outcomes

Pre-existing data sources include vital statistics (records of deaths, births, marriages etc), various medical and non-medical registries, as well as other health-relevant records (such as those maintained by schools, the armed forces, employers as well as by treating physicians). Additional cases may be found from established health surveys and screening programs. The advantages of using such pre-existing sources are that the cases are identified independently of twinship and zygosity, and participation in the twin/family study. Such data sources are often population based and systematic in data collection. On the other hand, most registers have not been designed primarily for research, meaning that there may be a great deal of within and between registry variability in the quality of the data. Finally, there may be limited access to the data because of technical, legal and ethical reasons. Research-oriented disease-specific registers do, however, exist for many conditions in many countries. The use of such a register is illustrated with data from the Finnish Cancer Registry.

Basal cell carcinomas of the skin

The cancer cases among the study subjects (see Box) in 1953–96 were identified by record linkage of the older, adult Finnish Twin Cohort data with the Finnish Cancer Registry data. The Finnish Cancer Registry is a nationwide database with information on all cases of cancer, diagnosed in Finland since 1953[12]. The reporting of cancer has been compulsory since 1961. Notifications on cancer patients are received independently from several sources such as hospitals, private physicians, and pathological laboratories. The cancer registry data can provide population incidence rates for comparison purposes. For example, in a study of the genetic epidemiology of basal cell carcinoma of the skin[13], we compared the incidence in twins with the population incidence data stratified on age, sex and calendar period. The standardized incidence ratio was not significantly different from unity, indicating that the overall incidence of BCC in twins was fully comparable to that in the population. Also, there were no differences by zygosity, sex or calendar period of follow-up.

Type I diabetes mellitus

Another approach was taken in a study of diabetes in twins[14], where no research-oriented register was available. Multiple sources of data were used to exhaustively identify all cases by combining all available data sources in which the goal was to identify cases of Type I and Type II diabetes in the Finnish Twin Cohort. The data sources were the register of in-patient hospital discharge diagnoses for 1972–85, diabetes medication records of the National Social Insurance Institution for 1964–1985 and cause of death data. These data records were linked to the twin data. Only a minority of subjects were found in both hospital and medication records, though all cases in death certificates were already in hospital records. Then, the paper records were retrieved and type of diabetes determined. The heritable component in Type I diabetes was estimated as 76%, compared to 75% in the Danish Twin Register study based on questionnaire screening, and clinical verification of case status[15]. Another advantage of using multiple data sources is that so-called capture-recapture methods[16] can be used to assess coverage, and provide a better estimate of the disease prevalence or incidence. This estimate can then be used in genetic modelling to provide a more accurate estimate of the heritable component in disease liability.

Collection of new data to identify outcomes

New data collection is necessary when existing data sources are inadequate for the study at hand. Most existing disease registers in any one area are limited. Mortality data does not adequately cover diseases with low mortality, such as many musculo-skeletal disorders, while hospital admission registers are limited to the most severe cases or to those patients with other co-morbid conditions that in combination lead to hospital care. Nearly all population-based epidemiological surveys are of unrelated individuals, and only household sampling will permit families to be studied. Birth cohort-based studies will pick up twin pairs, but since the twinning rate is generally around 1%, the birth cohort will have relatively few twin pairs in all but the largest studies.

After identifying the sample to be studied, the main methods of data collection are questionnaires, telephone interviews, and personal interviews, the last possibly combined with a clinical assessment and laboratory studies. Each technique has specific advantages and disadvantages. All have the disadvantage that some selection bias will always arise because not all subjects invited to participate will do so, and this is not a random process[9, 17]. Because twin pairs in which both twins participate are the most informative in twin studies, and the pairwise participation rate will always be at most equal, and generally somewhat lower than the rate computed for individual twins, twin studies need to aim for very high participation rates. For example, an individual participation rate of 85%, – considered excellent for most epidemiological purposes – would imply a pairwise rate of 72%, presuming that twins reply independently of their co-twin.

Questionnaires are inexpensive to administer compared to interviews because they can be centrally mailed to the subjects. Careful attention to the planning of the content and layout needs to be made, and questionnaires should always be tested in pilot samples. Questionnaires are impersonal, so social desirability bias associated with interviews is reduced. Because the investigator has no control over the situation when the questionnaire is being filled out, recognized disadvantages are that the wrong person may reply, and fairly high rates of missing data can be encountered. In twin and family studies, questionnaires may be mixed by, say, adolescent twins living in the same household. Also, collusion between family members in replying to certain items or entire sections of the questionnaire may occur. For cross-sectional, one-time surveys, it does not matter if the twins erroneously reply to each other's questionnaires, but for longitudinal studies this is a major concern. We have used the twins' forenames for them to identify the questionnaire correctly, if they live in the same household. Together with study specific codes, this permits identification of subjects while retaining the anonymity of individual questionnaires.

There is sometimes concern about the accuracy of self-reported medical history compared to medical records. This has been found to be disorder-specific. For example, a recent Finnish study of middle-aged adults found that self-report in questionnaires for common medical conditions found a very good level of agreement between questionnaire and medical records for cardiovascular conditions, but less adequate for musculo-skeletal disorders.[18]

Personal interviews have a number of distinct advantages that make them the method of choice in many disorders. Firstly, the identity of respondent can be verified, and outsiders can be excluded from the interview situation. Interviews are flexible, permit complex question patterns to be included, and allow probing to ensure that the questions are understood, and answers are unambiguous. Interviews as such are expensive and time-consuming. The quality control of interviews needs to be maintained through the data-collection period, with continuous assessment of inter- and intra-interviewer variability. Highly structured interviews or computer-assisted techniques alleviate this variability to some degree. Computer-assisted techniques help check for logical errors in data during the interviews, permit more complex question structures, and speed up the analyses, as the data becomes available 'online'. If the same interviewer always interviews both twins in a pair, which may be the case if an interviewer travels to see the twins, interviewer variability will be confounded with familial effects on the trait under study.

The third main method of data collection is one gaining in popularity as rates of telephone usage are increasing, even though unlisted numbers and answering machines may make it hard to reach subjects even in populations with universal telephone service. The advantage of **telephone interviews** is that they are cheaper than personal interviews, while retaining the flexibility of personal interviews. Also lesser interviewer bias may be found. A difficulty of telephone interviews is the problem of long or complex response options, which are hard to transmit reliably to the respondent by telephone; premailed response option forms is one approach to this problem.[17] Like in questionnaire studies, the identity of respondent may be difficult to verify for certain, and this is particularly true for family members, including twins, living together. Also, there is a need to ensure confidentiality of responses.

Internet survey studies have been piloted, but genetic epidemiology studies on twins based on Internet contacts have to our knowledge not been carried out. Twins with ready access to the Internet are likely to represent a selected fraction of the population, but an Internet survey could provide a new means to carry out cross-national comparisons of twin similarity. This might include subjects also from countries, where more traditional twin research groups are weakly represented.

While questionnaires and interviews can be used as diagnostic instruments by themselves, this is not always sufficient. However, all these techniques can be used to screen for potential cases, who may then be invited to participate in diagnostic **clinical, laboratory and imaging studies**, with corresponding savings in time and resources compared to studying all twins in the sample with the expensive techniques. The disadvantage of doing so is that any effective screening method will leave some cases unidentified. The screening method should have a high negative predictive power, i.e. that those persons with a negative result on screening should not turn out to have been true cases. At the same time, only a reasonable number of false-positives, who turn out not to be cases on further diagnostic testing, can be tolerated.

The art of outcome definition

Incidence (new onset) data is to be preferred over prevalence (existing) data for research on etiological factors in disease. Compared to prevalence data, incidence data is not confounded by factors affecting duration of disease, its natural course, treatment or recovery. Likewise, migration of cases in or out of the study area does not affect estimates based on incident cases. A further reason for using incident cases is that our analyses of genetically informative data are rarely only univariate. As the values of covariates (risk factors) may change after disease onset, these should be assessed before disease onset. Prevalence data is, nonetheless, often used because incidence data takes a long time to accumulate. Collaborative, multi-centre studies are often carried out to shorten an otherwise long period of follow-up. Diagnostic criteria vary over time, which needs to be taken into account in coding and analysis of case information. For example, during the duration of the Finnish Twin Cohort studies, ICD revisions 8,9 & 10 have been in use in Finland. While the gene pool changes slowly, environmental determinants can change rather rapidly. An example of this is the rapid decrease in levels of coronary disease risk factors, particularly smoking, seen recently in middle-aged men in many countries, including Finland.

The difference in concordance due to the use of prospective versus retrospective data is illustrated in Table 3.2 as an exercise with data on breast cancer concordance from the Finnish Twin Cohort based on cases diagnosed in 1976–1996. There was a different concordance for disease in MZ and DZ pairs depending on whether the perspective is prospective (assessment from 1976 onwards and based on all pairs alive in 1976) or retrospective (assessment from 1997 backwards and based on cases among pairs with both co-twins alive at the end of 1996).

In this example the concordances are lower in the retrospective assessment, and the sample size is also considerably smaller.

Table 3.2 – Difference in concordance due to prospective versus retrospective perspective on breast cancer diagnosed between 1976 and 1996 in female twin pairs from the adult Finnish Twin Cohort.

	Number of concordant pairs	Number of discordant pairs	Probandwise concordance
Prospective Pairs alive in 1976 (N = 6588)			
MZ	3	62	0.089
DZ	7	188	0.069
Retrospective Pairs alive at start of 1997 (N = 5343)			
MZ	2	34	0.056
DZ	5	120	0.040

CONCLUSIONS

Some issues in sample selection and outcome definition that are particular to twin studies were reviewed and illustrated in this chapter. However, in general the procedures regarding these issues in twin and twin-family studies are the same as those used in other epidemiological, in particular genetic epidemiological research. In research, random errors can be diminished relatively easily by increasing sample size, which is thus more a matter of allocating sufficient resources. Careful consideration of sample selection and outcome definition is necessary during the planning stages of a new study, because various biases can otherwise be introduced. It is very difficult to assess the effect of such biases, and controlling for such biases later during the analysis of data is generally difficult, if not impossible.

REFERENCE LIST

1 Neale MC, Cardon LR. *Methodology for Genetic Studies of Twins and Families.* Dordrecht: Kluwer Academic, 1992; 1–496

2 Kaprio J, Koskenvuo M, Rose RJ. Population-based twin registries: illustrative applications in genetic epidemiology and behavioral genetics from the Finnish Twin Cohort Study. *Acta Genet.Med.Gemellol.(Roma).* 1990;**39**: 427–439.

3 Phillips DIW. Twin studies in medical research: can they tell us whether diseases are genetically determined? *Lancet* 1993;**341** : 1008–1009.

4 Kaprio J, Sarna S, Koskenvuo M, Rantasalo I. The Finnish Twin Registry: formation and compilation, questionnaire study, zygosity determination procedures, and research program. *Prog.Clin.Biol.Res.* 1978;**24 Pt B**: 179–184.

5 Sarna S, Kaprio J, Sistonen P, Koskenvuo M. Diagnosis of twin zygosity by mailed questionnaire. *Hum.Hered.* 1978;**28**: 241–254.

6 Kaprio J. General principles of twin studies – applications to research on migraine. In: Olesen J, Bousser MG, eds. *Genetics of Headache Disorders.* New York: Raven Press, 1998 (in press)

7 Marenberg ME, Risch N, Berkman LF, Floderus B, Defaire U. Genetic Susceptibility to Death from Coronary Heart Disease in a Study of Twins. *N.Engl.J.Med.* 1994;**330**: 1041–1046.

8 Streiner DL, Norman GR. *Health measurement scales. A practical guide to their development and use.* Oxford: Oxford University Press, 1989;1–175

9 Norell SE. *A short course in epidemiology.* New York: Raven Press, 1992;1–192

10 Rothman KJ. *Modern Epidemiology.* Boston: Little, Brown, 1986;1–358

11 Khoury MJ, Beaty TH, Cohen BH. *Fundamentals of Genetic Epidemiology.* New York: Oxford University Press, 1993;1–383

12 Finnish Cancer Registry. *Cancer incidence in Finland 1995. Cancer Statistics of the National Research and Development Centre for Welfare and Health*. Helsinki: Cancer Society of Finland publication No. 58, 1997

13 Milan T, Kaprio J, Verkasalo PK, Jansén CT, Teppo L, Koskenvuo M. Hereditary factors in basal cell carcinoma of the skin: a population-based cohort study in twins. *Br.J.Cancer* 1998; 78: 1516–1520

14 Kaprio J, Tuomilehto J, Koskenvuo M, et al. Concordance for Type 1(insulin-dependent) and Type 2 (non-insulin-dependent) diabetes mellitus in a population-based cohort of twins in Finland. *Diabetologia* 1992;**35**: 1060–1067.

15 Kyvik KO, Green A, Beck-Nielsen H. Concordance rates of insulin dependent diabetes mellitus: a population based study of young Danish twins *B.M.J.* 1995;**311**: 913–917.

16 Hook EB, Regal RR. Capture-Recapture methods in epidemiology: Methods and limitations. *Epidemiol.Rev.* 1995;**17**: 243–264.

17 Armstrong BK, White E, Saracci R. *Principles of exposure measurement in epidemiology*. Oxford: Oxford University Press, 1995;1–351

18 Haapanen N, Miilunpalo S, Pasanen M, Oja P, Vuori I. Agreement between questionnaire data and medical records of chronic diseases in middle-aged and elderly Finnish men and women. *Am.J.Epidemiol.* 1997;**145**: 762–769.

4

PRACTICAL APPROACHES TO ACCOUNT FOR BIAS AND CONFOUNDING IN TWIN DATA

Alex J. MacGregor

ABSTRACT

To derive a valid assessment of genetic risk from twin data requires that the influence of potential sources of bias is taken into account. The principles underlying the approach to dealing with bias and confounding are well developed in standard epidemiological studies of risk factors for disease. Their application to twin data is, however, often not straightforward.

Bias may arise from sample selection, selective loss of information from the sample, and confounding. The former two are built in to the structure of the data. The investigator's role is to recognise their existence and assess their potential influence. In some instances specific analytical strategies may be needed to take into account the structure of the data to minimise bias. Such methods include those which account for the mode of ascertainment, for the presence of missing data from co-twins, and for time censored observations. Confounding may be addressed explicitly through analysis. Simple stratified analysis is often precluded by the complexity of twin data and modelling provides a more appropriate approach. The types of models employed include those based on linear and logistic regression, and variance components models using either the residuals of a regression analysis, variance decomposition techniques or full multivariate normal modelling approaches. The appropriateness of each of these methods depends on the biological question that is being addressed.

Caution needs to be exercised not to apply modelling approaches uncritically. Of particular concern is the lack of power to assess the influence of the shared environment, the use of the parsimony principle in model selection, and the use of statistical tests as a basis for selecting confounders. Used injudiciously, statistical modelling methods risk inflating precision and themselves biasing inference.

To the epidemiologist, the principal objective of collecting twin and family data is to assess whether familial aggregation of a disease or trait can be explained by common genetic or common environmental factors. If there is a genetic contribution, the strength of the genetic risk both to individuals and in populations also needs to be assessed.

Determining whether an inferred genetic influence is real or is the result of bias is an essential part of the analytical process, and is the subject of this chapter. Numerous types of bias are recognised in twin and family data and the analytical approaches developed to accommodate their effects are diverse and often specialised. In some circumstances different approaches have been used to tackle the same analytical problem and the 'correct' approach is less than clear.

This account focuses on the methods that have been used in accounting for the effects of bias and confounding in twin data in both studies of disease and of quantitative traits. The topics covered are wide-ranging and each could be the subject of detailed review in its own right. This treatment, however, is presented as an overview of the main issues and is aimed at practitioners rather than theoreticians. Some aspects are considered in greater detail in other chapters in this book. Chapters 2,3 & 6 by Strachan, Kaprio et al and Kyvik consider the specific effects of selection bias in twin data. Issues involved in modelling in general, and in variance component and survival analysis in particular, are given detailed treatment in chapters 9,10,11,13 & 18 by Christian, Snieder, Pickles, Hopper and Neale.

MEASURES OF GENETIC RISK

It is first useful to provide a broad definition of the genetic risks estimated from twin data on which the forces of bias and confounding may be acting (Box 4.1). Epidemiologists are familiar with the concept of quantifying the association between exposure and disease through estimation of risk ratios or risk differences. The concepts of 'exposure' and 'outcome' have been extended to twin and family data by analogy to conventional epidemiological study designs[1]. One view is to consider cross-sectional twin data as a 'reconstructed' cohort study design and to focus on the co-twins of affected twins[2]. In this design, the 'cases' are those co-twins of affected twins who develop incident disease and the 'exposure' is the degree of relationship (in this case zygosity) with the proband. This analogy can be extended to include other groups unexposed to the same genes, such as spouses of affected cases and matched controls selected from the general population. Using this approach, appropriate measures of 'genetic risk' include the ratio of concordance in MZ and DZ twin pairs and the difference in concordance between MZ and DZ twin pairs. Another measure is equivalent to the 'sibling recurrence risk'[3]: the ratio of the prevalence of disease in MZ and DZ co-twins of affected cases to the disease's population prevalence.

An alternative measure of 'genetic risk' used widely in variance components analysis of twin and family data is genetic variance. This is commonly summarised as the fraction of overall phenotypic variance attributed to genetic factors and expressed as 'heritability'[4]. Phenotypic variance can be measured directly for quantitative traits; for discontinuous traits (such as the presence or absence of disease)

BOX 4.1

Measuring genetic risk in twin epidemiology

CONCORDANCE

Concordance provides a measure of the proportion of co-twins of affected twins that have disease themselves.

Pairwise concordance (Cp)

This is the probability that a twin is affected, given that at least one member of the pair is affected. It is estimated by the ratio of concordant (C) to the sum of concordant (C) and discordant (D) pairs:

Cp $\qquad = C/C+D$

Casewise concordance (Cc)

This is the probability that a twin is affected, given that the co-twin is affected. It is estimated by the 'probandwise concordance estimator'. This includes a correction for dual ascertainment of concordant pairs under non-random (incomplete) sampling and takes into account the number of additional concordant pairs (C') ascertained independently through an affected twin[13]:

Cc $\qquad = C+C'/C+C'+D$

If an attempt is made to ascertain all affected twins in the population studied then

Cc $\qquad = 2C/2C+D$

GENETIC RISK

Genetic risk can be inferred from the comparison of estimates of concordance in MZ and DZ twins, or by comparison with the disease's population prevalence.

Concordance ratio: $\qquad = Cp_{MZ}/Cp_{DZ}$ or Cc_{MZ}/Cc_{DZ}

Concordance difference: $\qquad = Cp_{MZ}\text{-}Cp_{DZ}$ or $Cc_{MZ}\text{-}Cc_{DZ}$

Recurrence risks

Compare the risk of disease in a co-twin of an affected twin to the risk in the population (P).

MZ recurrence risk: : $\qquad = Cc_{MZ}/P$

DZ recurrence risk : $\qquad = Cc_{DZ}/P$

GENETIC VARIANCE

Analysis of variance methods based on the biometric model provide methods of explicitly estimating the extent to which phenotypic variation (Vp) can be attributed to population genetic variation[4]. Heritability is the ratio of genetic variation to overall phenotypic variation.

Phenotypic variance attributable to population genetic variance

$\qquad = Vg$

Heritability. $\qquad = Vg/Vp$

the variance can be inferred by considering the underlying liability to disease as being distributed on a continuous scale[4]. Conceptually, 'heritability' is similar in epidemiological terms to the notion of population attributable risk as it does not directly reflect the genetic risk to an individual. Heritability does *not* represent percentage of disease attributable to genetic factors[5] as it is sometimes naively interpreted.

SOURCES OF BIAS

The classic approach to considering the contribution of bias in epidemiological studies is to consider separately the effects of selection bias, information bias and confounding (Box 4.2)[6]. All have the potential to bias the measure of genetic risk estimated from twin data.

As well as sampling issues, assessing the existence and magnitude of bias requires an insight into the study design and the subject matter of the investigation and hence may require subjective judgements on the part of the investigator[7]. The use of statistics to assess the contribution of chance has no place in adjudicating on the presence or absence of bias in genetic epidemiological studies[6]. Instead, the contribution of analysis is: (i) to provide appropriate strategies for handling data from studies that are prone to introduce bias through their design and (ii) to provide consistent methods to adjust for the effects of bias. A number of these analytical strategies are discussed below.

BOX 4.2

Sources of bias in twin data

SELECTION BIAS

This occurs when there is a distortion in the effect measure that results from procedures used to select subjects. The potential for selection bias will always exist when the ascertainment of twins in a sample is not random.

INFORMATION BIAS

Results from errors in measuring a risk factor, and is therefore a ubiquitous component in all epidemiological studies. Examples include bias resulting form misclassification, inaccurate recall, and missing and censored data.

CONFOUNDING

Can be considered to be caused by 'mixing of effects'[6] and results in a distortion of the estimate of the size of risk associated with genetic 'exposure' by an extraneous factor. For all diseases with a complex genetic and environmental basis, the possibility of confounding needs to be considered.

SELECTION BIASES

Ascertainment

An optimal strategy for minimising extraneous biases in twin data would be to obtain a truly random population sample[8]. This approach is only practical, however, in certain situations, for example where diseases are common or where data are readily available from a population register. Most designs rely on systematic (non-random) sampling methods to obtain probands and the need for analysis to take into account sampling in interpretation of twin and family data has long been recognised[9,10].

Concordance measures

An elementary analytical problem relating to ascertainment is the question of the most appropriate way to express twin concordance: either pairwise or casewise (Box 4.1)[11]. Both measures of concordance refer to a conditional probability that a given twin is affected; for the former, conditioning is on at least one member of the twin pair being affected, while for the latter, conditioning is on the co-twin being affected. To an extent, it is up to the investigator to decide for themselves between these two measures; if results are presented consistently and in combination with raw data, there can be little confusion. Problems arise, however, if concordance results are compared between studies with different levels of ascertainment, and if it is the intention to use concordance as a comparative measure of risk to a relative of an affected case versus risk to an individual drawn at random from the population. These last two aims are better served by the use of casewise concordance, which is estimated by the 'probandwise concordance estimator'[11], that effectively augments both the numerator and the denominator by the number of concordant pairs ascertained independently through both affected twins[12,13]. The probandwise concordance estimator thus examines the risk to an individual in the dataset, taking into account the ascertainment procedure. The ratio of casewise concordance in DZ twins to population prevalence is equivalent to the sibling recurrence risk.

Non-random sampling

A number of well-described approaches have been developed to account for selection based on the ascertainment of affected individuals (probands)[8,10]. These provide an appropriate analytical structure to account for the potential bias that may be present in a naive analysis. Martin and Wilson[14] provide a mathematical approach to dealing with non-random sampling schemes in variance components analysis of twin data. The essence of their approach is that it divides the likelihood by the proportion of the population remaining in the study after ascertainment. Computationally, this method can be readily implemented in Mx[15].

More complex ascertainment biases

The effect on concordance of more complex forms of non-random ascertainment has been examined by Kendler and Eaves[16]. Two particular types of ascertainment were assessed: (i) concordance-dependent, where the probability of becoming a proband differs in affected members of concordant when compared with discordant pairs and (ii) concordance non-independent (or correlated) ascertainment, where the probability of becoming a proband in a concordant pair depends on the ascertainment status of the affected co-twin. Both biases are commonly encountered in twin studies. The increased burden of disease for a family with disease concordant when compared with disease discordant twin pairs, for example, may increase the chances of both members presenting for treatment (concordance dependent bias). The outcome of one twins' treatment experience may affect the chance that their affected co-twin will also come forward for treatment (correlated ascertainment bias).

Kendler and Eaves show that both methods of ascertainment have the potential to bias estimates of casewise concordance and heritability. In most situations, concordance-dependent ascertainment had the greater effect. The effect on the estimated casewise concordance and heritability was complex and determined by the relative probability of ascertainment in affected members of concordant and discordant pairs. The direction of bias on heritability was further determined by the population prevalence of the disease.

Kendler and Eaves' formulation provides an approach to 'adjusting' the effects of these sampling biases. This requires information on three probabilities: (i) the probability that an affected member of a concordant pair will be ascertained given that a co-twin has not been ascertained; (ii) the probability that an affected member of a concordant pair will be ascertained given that the co-twin has already been ascertained; and (iii) the probability that an affected member of a discordant pair will be ascertained. The strength of the effects demonstrated by their analysis makes a compelling case for collecting data whenever possible in a practical setting to allow these probabilities to be estimated.

Volunteer bias

The representativeness of the sample as a random sample of the general population is the chief concern in studies based on volunteers. Lykken et al[17] studied the results of eleven volunteer-based twin studies conducted between 1928 and 1977 and concluded that all the samples showed a consistent excess of female and monozygotic twins in a ratio of 2:1 (the so-called 'two thirds rule' for volunteer samples). The consequence of this disparity in ascertainment was that male and DZ pairs were less likely to be representative of their underlying populations because they were more highly selected. This was supported by a tendency for MZ and female pairs to show smaller between pair variance.

These observations led Lykken et al to propose that volunteer bias could be 'corrected' if it was assumed that the true total variance of MZ and DZ twins equalled the variance in a 'normal' group of non-twin non-volunteers ascertained from the general population. The 'corrected' between pair means square was obtained by subtracting the within pair mean square from twice the variance of the normal group. The rationale for this approach was that the correction did not favour either MZ or DZ twin types. Lykken et al examined the size of the bias in young adult twins recruited from the Minneapolis area and showed the correction to have a substantial influence on the assessment of the genetic influence on both anthropometric and behavioural variables. In general, DZ intraclass correlations were systematically overestimated in the uncorrected analysis.

The influence of volunteer bias on the interpretation of results from twin studies was further examined in later analysis by Lykken *et al.* using Monte-Carlo methods based on the Martin -Wilson model (which allows differential rates of twin partic- ipation) and an adaptation which accounts for a possible direct effect of intrapair similarity in the desire to participate[18]. Their conclusion was that volunteer bias had unpredictable effects, sometimes leading to large overestimates of twin correlation, and that the effects tended to be larger in MZ than DZ twins.

The formal mathematical models that have been developed to describe the action of volunteer bias provide, in theory at least, a means for adjusting for its effects. All approaches, however, require a degree of mathematical consistency in the processes involved that is difficult to verify in a practical setting. Analyses based on the models discussed earlier, however, stress to investigators the importance and potentially unpredictable nature of this type of bias on the interpretation of twin data and it is therefore critically important to be aware of its potential effects.

In assessing the importance of volunteer effects in a dataset Lykken et al emphasise the value of analysing and testing data for equality of means and variances of MZ and DZ twins, and comparing the variance in twins with that in the general popu- lation[18]. There is also a strong case for collecting adequate data from the sample (such as on variables related to the liability to volunteer) to allow the volunteer effects to be studied through appropriate modelling methods[19]. The importance of simple conventional methods, including obtaining whatever information is available on non-responders, also cannot be over-emphasised.

Other consequences of selection bias

Non-random ascertainment schemes clearly introduce the opportunity for gener- ating imbalance in samples for a large number of variables which may influence the interpretation of twin data. Imbalances are typical of most large twin datasets: for example, the Australian Twin Registry shows differences between MZ and DZ twins for age, gender, ethnicity, geographical location and socio-economic status[20]. If sampling is restricted to a geographical region, this in itself may introduce bias. This has been demonstrated in data from the Finnish Cohort[21] in which the cumulative concordance for psychosis was greater for MZ twins living in the same province than for MZ twins living further apart. Bias also arises when studies focus on particular occupational groups[22,23].

For analysis, variables found to be in imbalance through selection effects may be considered as potential confounders and appropriate methods for dealing with confounding in twin data are discussed later. It is clear, however that unless attention to study design is meticulous, the scope for selection bias is potentially wide.

INFORMATION BIAS

Misclassification and measurement error

In general, random measurement errors and misclassification would be expected to have a predictable effect on the measurement of genetic risk, biasing it towards the null. If there are systematic differences in the error rate between MZ and DZ and concordant and discordant pairs, however, there may be instances where the effect of information bias may be more complex. For example, recall may be more common for a concordant event, and parental ratings of concordance may be influenced by the perceived zygosity of twins. If the pattern of information bias is regular, however, modelling approaches can adjust for their effects.

Rater bias. Neale et al have developed structural models which adjust the effects of parent's rating of their children's phenotypes by allowing the latent phenotypes of children to be considered independently of rater biases and the unreliability of raters[24].

Recall bias. Models for explicitly assessing the extent of recall bias have been examined by Pickles et al[25]. These analyses have specifically examined the effect of 'telescoping' – where the recalled date of distant events is shifted more towards the present than the recalled date of recent events[26]. In their multivariate model, Pickles et al showed that telescoping made a significant contribution to the maternal report of age of menarche and age at onset of breast development in twin daughters. They showed that, if unaccounted for, this bias would have had a substantial impact on the estimate of genetic and environmental variance.

In many practical circumstances, the effect of measurement error and misclassification are less regular and hence cannot be modelled adequately. An example was encountered in the analysis of data from the ARC UK Nationwide Twin Study of Rheumatoid Arthritis[27] where a number of alternative schemes were available for classifying the disease. Under different classification schemes, the MZ/DZ concordance ratio for rheumatoid arthritis varied between 2 and 5 indicating a wide variation in genetic risk. The variation broadly reflected differences in the prevalence of disease captured by each scheme, with higher concordance ratios found where the disease was more prevalent. It was not, however, possible to predict the variation in concordance estimates entirely on this basis and it is possible that the schemes may have incorporated aspects of the clinical disease that had a different genetic contribution[28].

Missing data

Studies in which the unit of observation is a group of relatives are more susceptible to the effects of missing observations than those in which the unit of observation is a single individual. As with misclassification, the central issue with respect to bias is whether or not data are missing at random. In practice, there do not appear to be simple or standard approaches to handling missing data or guidelines for investigators regarding the extent of its effects[29]. Conventional statistical approaches for assessing the effects of missing data apply as equally to twin data as to any other dataset. These include deleting variables, interpolation, and treating missing data as dummy data. Neale discusses some practical aspects of handing missing data from twins in the structural modelling software Mx[15]. The software allows likelihood estimates to be based on raw data for each observation in a sample and thus provides a more efficient alternative to deleting subjects or pairs. Clearly, however, the method does not circumvent the problem of bias if there are systematic reasons for the absence of data.

Censored data

It is important to take into account information on 'exposure' to a genetic risk appropriately in analysis. An earlier mean age at disease onset in MZ when compared to DZ twins may, for example, lead to a spurious estimate of a genetic effect if concordance data from the twins are examined in cross sectional design. Using the analogy of a 'reconstructed cohort' design discussed earlier, in discordant pairs the absence of disease in the co-twin of an affected twin can be considered to be censored at the time that the pair was ascertained into the study. Similarly, unaffected pairs (if also ascertained) could be considered as being doubly censored. Survival approaches exist to analyse data of this type and have been applied to twin data[30–32]. Survival methods are considered in detail by Pickles in Chapter 11.

CONFOUNDING

Extending the classical epidemiological definition of confounding[6] to twin data is not straightforward. The central issue, however, is whether a variable distinct from the disease or trait of interest could give rise to a spurious association which may be attributed to genetic causes purely because of its imbalance in a dataset. It is possible to envisage at least two ways in which such an imbalance could bias an estimate of genetic risk. First, a variable associated with disease in its own right might show greater correlation in MZ than DZ twin pairs. An example might be the spurious suggestion of a genetic association occurring in a smoking-related disease resulting from a greater correlation in smoking status in MZ when compared with DZ twins. This type of confounding would occur even if the mean value of the 'confounding' variable was balanced in the MZ and DZ groups. Second, spurious genetic associations might arise if the level of the confounding variable affected the covariance of the trait within twin pairs[20]. Age affects the covariance of a number of traits including, for example, serum lipid levels[33]; an imbalance in age in MZ and DZ groups could thus give rise to a spurious genetic association.

In studies of exposure and disease outcome the effect of confounding can be dealt with analytically either by stratification or by modelling. The use of stratification to assess the effects of confounding has a number of appealing advantages. No assumptions are needed concerning the underlying distribution of confounding variables, or the relationship between effects measured in different strata. Stratum-specific estimates are readily interpretable, and provide a direct indication of interaction and effect modification. The chief limitation of the approach is the requirement for large numbers, especially when examining the effects of multiple confounding variables. This is especially pronounced when the method is applied to studying twin and family data. As a simple example, to stratify twins on smoking status, crudely classified as 'current' 'past' or 'never', requires nine categories of concordance and discordance in both MZ and DZ twin groups. Examining two or more confounding variables in this way rapidly becomes impractical.

In contrast to stratification, modelling approaches are better suited to accommodating sparse data and have become the mainstay of statistical approaches for dealing with confounding in twin and family data. The literature contains many approaches to modelling confounding that can be applied to twin data. Interestingly, however, there is little discussion of their relative merits and their potential pitfalls. There is little to guide investigators as to which approach to use in differing situations. As would be expected for methods attempting to achieve a common aim, many overlap both conceptually and mathematically. Several of these methods are summarised in Box 4.3 where they have arbitrarily been divided into methods that focus on the measurement and adjustment of absolute or relative risk to individuals or pairs (risk-based), and methods that focus on adjusting genetic variance and heritability (variance-based).

Differences in approaches

'Risk-based' approaches (Box 4.3) have the advantage that they produce an assessment of risk to an individual (for example the co-twin of an affected twin) or

BOX 4.3

Analytical approaches to confounding

RISK-BASED APPROACHES

The MZ/DZ concordance odds ratio. The logarithm of the concordance odds ratio can be modelled as a function of the matched covariates of co-twins of affected twins[1].

Logistic regression models. The probability of a co-twin of an affected twin developing disease themselves can be modelled by logistic regression. Including zygosity in the logistic model provides an estimate of genetic risk and a test of significance under the assumptions of the classical twin model. This can be conveniently adjusted for the co-twins covariates[1,34]

Survival methods. Cox proportional hazards modelling has been used to compare the concordance hazard in MZ versus DZ co-twins accounting for the differences in exposure interval (time to disease onset) in pairs. This method allows for the effects of covariates in MZ and DZ co-twins to be taken into account[27,43].

Multivariate methods. The logistic model can be extended to consider both twin and co-twin independently along with their covariates by the use of multiway frequency analysis (log-linear modelling)[1,35]. Continuous data can be modelled similarly using a generalised estimating equation based approach[44]. Bivariate survival models have also been extended to include covariates[30].

VARIANCE COMPONENTS APPROACHES

The DeFries-Fulker regression equation[45,46] allows a proband's score to be predicted from their co-twin's score. Terms in the model provide a direct estimate of heritability under the assumptions of the classical twin model, and the model can be extended to include covariates.

Variance components modelling. The full multivariate nature of twin data containing main effects and confounding variables in both twin and co-twin can be modelled through variance decomposition approaches[24] and in the multivariate normal model[36,37]. These approaches account for the possibility of both genetic and environmental variance shared between main effects and confounding variables.

'Restricted' variance components methods apply classical univariate variance components analysis to the residuals of a linear regression of the main effects against confounding variables[39].

relative risk (MZ versus DZ co-twin) that has a simple interpretation. They also do not require an explicit assumption about the similarity of the shared common environment in MZ and DZ twins[34,35]. They are suited to the analysis of discontinuous data and do not rely on a need for the assumption that there is an underlying liability that is normally distributed[35]. They do not, however, provide a test of specific genetic or environmental models that underlie familial aggregation[34].

Variance-based methods can be applied equally to continuous data, to discontinuous data (through transformation to a normal liability distribution) and to mixtures of data types. Variance decomposition methods (including the use of the Cholesky decomposition[24]) allow specification of a full model that provides a complete description of the variance structure of the data. Submodels may be explored to provide further insight into the genetic and environment structure of the data and the nature of confounding. The multivariate normal model provides a similarly complete approach to modelling both the mean and variance structure of complex genetically informative datasets[36,37]. These methods are readily implemented using software including Mx[15] and FISHER[38]. The chief limitation of these approaches is that they are laden with assumptions including a requirement for multivariate normality and the absence of interaction effects. Many of these are impossible to test[1,35]. Furthermore, for datasets encountered in many practical situations with several covariates including data that is not normally distributed and ordinal data, the solution to the likelihood equations may be found to be numerically intractable.

The 'restricted variance components' approach[39] appears to have all the disadvantages of the multivariate approach in terms of its dependence on assumptions. It also does not allow an assessment of shared genetic and shared environmental influence between confounders and main effects. The precision of the estimates of variance components derived using this approach may also in some circumstances be inaccurate. In practical terms, however, the method is simple to execute and is applicable to datasets involving large numbers of potential confounders. If the main aim of the analysis is to provide reassurance that confounding does not influence estimates of the absolute size of genetic variance, this approach provides a widely applicable and straightforward solution.

FURTHER CONSIDERATIONS

Several aspects of the analytical treatment of bias, in particular the treatment of confounding in twin data, merit further discussion.

The validity of 'adjusting' genetic risk

Extending the analogy of confounding in epidemiological studies estimating the relationship between 'exposure' and 'outcome' to studies estimating genetic risk gives rise to conceptual difficulties. If confounding is the result of genetic factors shared with an extraneous variable, is it appropriate to effectively remove the influence of these shared genetic factors when assessing the overall genetic contribution to disease? The same question arises with respect to shared environmental influences. If the objective is to assess genetic and environmental effects independently of all other possible aetiological factors, it may be appropriate to 'adjust'. The process of 'adjusting' may well, however, result in an estimate of a different entity. The heritability of a measure is always estimated in the context of its mean; each time one uses different factors to model the mean, one is actually dealing with a different trait[40].

Heritability is a measure of the extent to which variation in a trait is attributable to genetic variance in the population as a whole and is not a direct measure of risk to an individual. Fisher himself pointed out that the denominator of heritability (the phenotypic variance) is a 'hotch potch'[20]. As well as genetic factors, environmental effects, measurement error and pure error all contribute. It is self-evident that any procedure that 'adjusts' for the effects of known environmental variables and for measurement error is always likely to increase an estimate of heritability. By including more confounding variables and providing a better model for the data, the precision of estimates of the genetic effect may increase. This contrasts with the effect of including confounding variables in conventional case-control studies of exposure and outcome. There is considerable potential, therefore, to provide false reassurance about the validity of genetic effects through injudicious analyses focusing on heritability alone.

Model selection

The rationale for choosing which potential confounders, or covariates, to include in a model and finding an appropriate model for the data is a further point of discussion. Traditional modelling approaches often rely on a balance between the goodness of fit of a model and the model's parsimony, as assessed by the numbers of parameters, for the choice of the most suitable model to describe the data. This appealing strategy has its origins in the views of the 14th century nominalist philosopher and theologian William of Occam, and has been embraced seemingly without questioning by quantitative geneticists[41]. This approach however appears to be unjustified in assessing the effects of bias and confounding. In judging the importance of a confounding variable, it is the influence of the confounder on the estimated size of the measure of effect that is relevant rather than the statistical significance of its contribution to a model[6]. In Chapter 13, Hopper points out that the use of the 'parsimony principle' tends to lead to rejection of the common environmental effects in all but the largest of twin datasets. Likewise, as the effects of individual confounding variables on genetic and environmental variance is likely to be small[42], the approach favours rejecting potential confounding variables from a model. Careless application of parsimony and goodness of fit in modelling runs the risk of itself introducing bias, as well as falsely inflating statistical precision.

An epidemiological approach would be to favour including all relevant potential confounders in all models on the basis of an *a priori* judgement of their contribution[6]. In variance components analysis, which allows common environmental effects to be estimated explicitly, a logical extension of this view would be to recommend that investigators always at least report parameter estimates from models including the common environmental variance, even if they are negative. Standard errors, confidence intervals or some other measure of the imprecision of the estimates should be quoted.

CONCLUSION

Estimates of genetic risk are inevitably influenced to a greater or lesser extent by the effect of bias and confounding in all studies of twins. Much effort can be made to minimise bias through meticulous attention to all aspects of study design. However, it is also critical that investigators collect data items that allow an assessment of the importance of bias through quantitative means. In assessing the effect of confounding in a dataset it is important to have a clear concept of the measure of effect that is of interest and of the relevance of taking into account the effects of extraneous variables. No single method for accounting for confounding variables appears to be ideal, however the approach to handling confounding should be dictated by epidemiological considerations, and not by simply providing the 'best fitting' model.

Acknowledgements

I am grateful to John Hopper for his comments on this manuscript.

This work has been supported by a grant from the Arthritis Research Campaign

REFERENCES

1. Khoury MJ, Beaty TH. Fundamentals of Genetic Epidemiology. New York: Oxford University Press, 1993.

2. Susser E, Susser M. Familial aggregation studies. A note on their epidemiologic properties. Am J Epidemiol 1989; 129(1): 23–30.

3. Risch N. Linkage strategies for genetically complex traits. Am J Hum Genet 1990; 46: 222–8.

4. Falconer DS. Introduction to quantitative genetics. Harlow: Longman Scientific and Technical, 1989.

5. Hopper JL. Heritability. In: Armitage P, Colton T, Eds. Encyclopaedia of Biostatistics. New York: Wiley, 1998: 1905–6.

6. Rothman KJ. Modern Epidemiology. Boston: Little, Brown and Company, 1986.

7. Greenland S. Response and follow-up bias in cohort studies. Am J Epidemiol 1977; 106: 184–7.

8. Hodge SE. Ascertainment. In: Armitage P, Colton T, Eds. Encyclopaedia of Biostatistics. New York: Wiley, 1998: 197–201.

9. Fisher RA. The effect of methods of ascertainment upon the estimation of frequencies. Ann Eugen 1934; 6: 13–25.

10. Morton NE. Outline of Genetic Epidemiology. Basel: Karger, 1982.

11. Hopper JL. Twin concordance. In: Armitage P, Colton T, Eds. Encyclopaedia of Biostatistics. New York: Wiley, 1998: 4626–9.

12. Allen G, Harvald B, Shields J. Measures of twin concordance. Acta Genet.(Basel.) 1967; 17: 475–81.

13. Emery AH. Methodology in medical genetics: an introduction to statistical methods. Edinburgh: Churchill Livingstone, 1976.

14. Martin NG, Wilson SR. Bias in the estimation of heritability from truncated samples of twins. Behav Genet 1982; 12: 467–72.

15. Neale MC. Mx: Statistical modeling. 4th Edition. Box 126 MCV, Richmond, VA 23298: Department of Psychiatry. 1997.

16. Kendler KS, Eaves LJ. The estimation of probandwise concordance in twins: the effect of unequal ascertainment. Acta Genet Med Gemellol (Roma) 1989; 38: 253–70.

17. Lykken DT, Tellegen A, DeRubeis R. Volunteer bias in twin research: the rule of two thirds. Soc.Biol. 1978; 25: 34577.

18. Lykken DT, McGue M, Tellgren A. Recruitment bias in twin research: the rule of two thirds reconsidered. Behav Genet 1988; 17: 343–62.

19. Neale MC, Eaves LJ. Estimating and controlling for the effect of volunteer bias with pairs of relatives. Behav Genet 1993; 23(3): 271–7.

20. Hopper JL. The epidemiology of genetic epidemiology. Acta Genet Med Gemellol (Roma) 1992; 41(4): 261–73.

21. Romanov K, Koskenvuo M, Kaprio J, Sarna S, Heikkilè K. Selection bias in disase-related twin studies. Acta Genet Med Gemellol (Roma) 1990; 39: 441–6.

22. McMichael AJ. Standardised mortality ratios and the 'healthy worker effect': scratching beneath the surface. J Occup Med 1976; 18: 165–8.

23. Reed T, Quiroga J, Selby JV et al. Concordance of ischemic heart disease in the NHLBI twin study after 14- 18 years of follow-up. J Clin Epidemiol 1991; 44(8): 797–805.

24. Neale MC, Cardon LR. Methodology for genetic studies in twins and families. Dordrecht: Kluwer Academic Publishers, 1992.

25. Pickles A, Pickering K, Simonoff E, Silberg J, Meyer J, Maes H. Genetic 'clocks' and 'soft' events: a twin model for pubertal development and other recalled sequences of developmental milestones, transitions, or ages at onset. Behav Genet 1998; 28(4): 243–53.

26. Sudman S, Bradburn NM. Effects of time and memory factors on response in surveys. J Am Stat Assoc 1973; 63: 805–15.

27. Silman AJ, MacGregor AJ, Thomson W et al. Twin concordance rates for rheumatoid arthritis: results from a nationwide study. Br J Rheumatol 1993; 32(10): 903–7.

28. MacGregor AJ, Bamber S, Silman AJ. A comparison of the performance of different methods of disease classification for rheumatoid arthritis. Results of an analysis from a nationwide twin study. J Rheumatol 1994; 21(8): 1420–6.

29. Tabachnick BG, Fidell LS. Using multivariate statistics. New York: HarperCollins College Publishers, 1996.

30. Clayton DG. A model for association of bivariate life-tables and its application in epidemiological studies of familial tendency in chronic disease incidence. Biometrika. 1978; 65: 141–51.

31. Pickles A, Neale M, Simonoff E et al. A simple method for censored age-of-onset data subject to recall bias: mothers' reports of age of puberty in male twins. Behav Genet 1994; 24(5): 457–68.

32. Pickles A, Crouchley R, Siminoff E, Eaves L, Meyer J. Survival models for developmental genetic data: age of onset of puberty and antisocial behaviour in twins. Genet Epidemiol 1994; 11: 155–70.

33. Snieder H, van Doornen LJ, Boomsma DI. The age dependency of gene expression for plasma lipids, lipoproteins, and apolipoproteins. Am J Hum Genet 1997; 60(3): 638–50.

34. Ramakrishnan V, Goldberg J, Henderson WG et al. Elementary methods for the analsysis of dichotomous outcomes in unselected samples of twins. Genet Epidemiol 1992; 9: 273–87.

35. Hopper JL, Hannah MC, Macaskill GT, Mathews JD. Twin concordance for a binary trait III A bivariate analysis of hay fever and asthma. Genet Epidemiol 1990; 7: 277–89.

36. Hopper JL, Mathews JD. Extensions to multivariate normal models for pedigree analysis. Ann Hum Genet 1982; 46 (4): 373–83.

37. Lange K, Westlake J, Spence MA. Extensions to pedigree analysis. III. Variance components by the scoring method. Ann Hum Genet 1976; 39(4): 485–91.

38. Hopper JL. Review of FISHER. Genet Epidemiol 1988; 5(6): 473–6.

39. Williams CJ, Christian JC, Norton JAJ. TWINAN90: A FORTRAN program for conducting ANOVA-based and likelihood- based analyses of twin data. Comput Methods Programs Biomed 1992; 38: 167–76.

40. Hopper JL. Variance components for statistical genetics: applications in medical research to characteristics related to human diseases and health. Stat Methods Med Res 1993; 2(3): 199–223.

41. Mulaik SA, James LR, Van Alstine J, Bennett N, Lind S, Stilwell CD. Evaluation of goodness-of-fit indices for stuctural equation models. Psychol Bull 1989; 105(3): 430–45.

42. Khoury MJ, Beaty TH, Liang KY. Can familial aggregation of disease be explained by familial aggregation of environmental risk factors? Am J Epidemiol 1988; 127(3): 674–83.

43. Snieder H, MacGregor AJ, Spector TD. Genes control the cessation of a woman's reproductive life: a twin study of hysterectomy and age at menopause. J Clin Endocrinol Metab 1998; 83(6): 1875–80.

44. Liang K-Y, Zeger SL. Longitudinal analysis using generalized linear models. Biometrika 1986; 73: 13–22.

45. DeFries JC, Fulker DW. Multiple regression analysis of twin data. Behav Genet 1985; 15: 467–73.

46. DeFries JC, Fulker DW. Multiple regression analysis of twin data: etiology of deviant scores versus individual differences. Acta Genet Med Gemellol (Roma) 1988; 37: 205–16.

5

THE CO-TWIN CONTROL STUDY

David L. Duffy

ABSTRACT

Twins have been used as co-twin control studies since the era of Galton to examine the effects of the environment. One of the original purposes in setting up the large Scandinavian twin registries was to collect disease and exposure discordant twin pairs, allowing an analysis of environmental risks of disease, whilst controlling the effects of genetics by matching. This was partly motivated by the "smoking controversy" of the 1950s, where RA Fisher and others suggested that gene-environment (here a risk behaviour) correlation might be a confounding factor in conventional case-control and indeed cohort studies.

Although the analysis and interpretation of such studies of identical twins is straightforward, combining results from identical and nonidentical twin pairs using path models allows additional inference about genetic and environmental contributions to the trait-covariate association. This design has advantages and disadvantages compared to both traditional case-control and twin studies.

Cotwin case control studies can also be used to examine gene-environment interactions and gene-gene interactions. In general these studies are useful whenever studying a heritable trait and can add considerable power when examining for G x E and when phenotype measurements are expensive.

Along with the classical twin design comparing monozygotic (MZ) and dizygotic (DZ) twins, the co-twin control study seems to have been a common concept with no particular originator. Galton's original method had been to examine twin concordance versus differences in life history.[1] He noted differences in personality (his area of interest) between twins could arise from a single episode of illness affecting one but not the other twin.[2] In a related vein, Newman, in the 1920s and 1930s, described life histories of MZ twins reared apart, looking for discordant outcomes that particular features of upbringing might explain.[3]

THE CO-TWIN CONTROL STUDY

The experimental co-twin control study

Newman et al also[3] described a co-twin control study. In this design, one member of an MZ twin pair underwent an experimental treatment, and was compared in a matched fashion to the control co-twin. This methodology was also being explored by Gesell at roughly the same time.[4] A large number of such matched-design experiments have been carried out. The paper by Christian and Kang [5] gives calculations for the relative efficiency (RE) of the use of MZ twins versus unrelated subjects. They show, for example, that a trial of a lipid-lowering drug requiring 24 unrelated individuals (12 treated, 12 controls) would require only three sets of MZ twins (RE=4). For a trait under simple additive genetic control (no effects of family environment, genetic dominance or interaction) measured perfectly reliably this is (asymptotically):

$$RE = \frac{h^2}{1-h^2}$$

where h^2 is the heritability.

The observational co-twin control study

A related observational design, also known as a co-twin control study, is the longitudinal study of a cohort of twin pairs discordant in level of exposure to a putative disease risk factor. Occurrence of the disease in the followup period is recorded, and a matched analysis used to estimate the relative risk of the disease for a given exposure.

The co-twin case-control study

A final type of design is simply the 1:1 matched case-control study applied to twin pairs discordant for the outcome of interest. This may be nested within a twin-based cohort study. This type of study is slightly less common in the literature.

MOTIVATION FOR THE CO-TWIN CONTROL DESIGN

The purpose of any type of matched design is to control confounding. That is, the investigator believes variation in the trait of interest has a number of proximate causes or correlates, which may interact in a complex manner. To isolate the effects of the one covariable of current interest, the other factors are to be "held constant".

In the co-twin control design, the trait of interest is under genetic control, genes being the nuisance variables to be controlled. There are three broad patterns that might govern the relationship between trait, covariable and genes.

Firstly, the correlation between trait and covariable might be environmental in nature, where we wish to eliminate "noise" due to the genetic causes of the trait. The genes underlying the trait are unmeasured, so some type of family design is needed to indirectly assess them. If the trait heritability is high, a matched design will be highly efficient.

Alternatively, the correlation between trait and covariable is genetic in nature, in which case there is genetic confounding. Fisher's hypothesis that a particular genotype might increase both the propensity to smoke cigarettes and develop lung cancer is an example of this.[6]

The third possibility is that the correlation between trait and covariable is genetic or environmental, but there is interaction between the covariable and (other) genetic causes of (variation in) the trait -- gene by environment (G×E) or gene by gene (epistatic, G×G) interaction.

SOME EXAMPLES

To make matters concrete, I will present examples of the latter two epidemiological designs. A Medline search using the single keyword "co-twin" finds 50 studies since 1990, while there were at least the same number again when this search was widened to "twin".

Observational co-twin control studies

(a) Lassila et al[7] studied nine MZ male twin pairs discordant for smoking. The smokers had imbibed a mean of 14.7 pack-years, and did not differ from their non-smoking co-twins in height, serum levels of glucose, total cholesterol, LDL, or creatinine clearance. On duplex Doppler ultrasonography, the mean atherosclerotic plaque area in the carotid arteries of the smokers was 6.5 mm²; in the nonsmokers, 1.9 mm²; mean paired difference 95% confidence interval, 0.9–8.3 mm².

(b) Floderus et al[8] examined deaths due to coronary heart disease and cancer in the Swedish Twin Registry. A total of 10945 twin pairs (born 1886-1925) who returned a 1961 questionnaire reporting smoking habits were followed to 1981 (6447 deaths). A nested sample of smoking-discordant pairs were examined (Table 5.1).

The point estimates from the co-twin study agree with those from a standard analysis of the entire cohort (men 1.4, 95%CI=1.3-1.5; women 1.4, 95%CI=1.3-1.5)

(c) Hopper and Seeman[9] compared lumbar spine and femoral neck bone density among 21 MZ female and 20 DZ female twin pairs discordant for level of smoking. The lumbar spine density was on average 5.0% lower in the heavier smoker than their co-twin (SE=2.0%), increasing to a 9.3% difference in the 20 pairs where the difference in smoking level was greatest. This effect was not mediated by exercise level, coffee intake or body weight.

The co-twin case-control study

(a) Cicuttini et al[10] examined female twins discordant for radiographic signs of osteoarthritis (OA) at different joints. For 70 MZ and DZ twin pairs discordant for tibio-femoral joint (knee) OA, the affected twin was on average 2.9 kg heavier than

Table 5.1 – Mortality in smoking-discordant twin pairs born 1886–1925. Adapted from Table 3 of Ref 8.

Type of Twin pair	Number of discordant pairs	Smoker died first	Non-smoker died first	Relative Risk (95% CL)
MZ Males	133	35	22	1.6 (0.9–2.8)
DZ Males	413	102	72	1.4 (1.0–1.9)
MZ Females	224	37	32	1.2 (0.7–1.9)
DZ Females	595	105	83	1.3 (0.9–1.7)

the unaffected co-twin (95%CI=0.9–4.9 kg). No marked differences between MZ and DZ groups were detected, and the effect size estimate agrees with those from cohort and cross-sectional studies.

(b) Vagero and Leon[11] performed a nested study in the Swedish Twin Register cohort looking at effects of adult height on risk of death due to coronary heart disease. They ascertained 1394 pairs where one twin had died with a diagnosis of coronary artery disease. The shorter twin (difference 1 cm or greater) was found to have an elevated risk of CHD (1.15, 95%CI=1.03-1.25). There was evidence of a dose-response effect on conditional logistic regression analysis of the data.

(c) Swerdlow et al[12] compared risk of various cancers versus birth order in twins. Cancer registry data was linked to the National Health Service Central Register to identify twin births. There was no global trend, but for testicular tumours (Table 5.2), they observed an interaction between sex of co-twin and birth-order.

I suspect this is probably a Type I error, in view of the large number of cancers the authors examined.

A PATH MODEL FOR THE CO-TWIN CONTROL STUDY

A problem in interpreting data from MZ and DZ twins

In the examples above, estimates have usually been presented for two (or more) different zygosity groups. It seems logical that the degree of matching on genotype is different in these different groups, and so combining them may be sometimes less than straightforward.

Below (Table 5.3) are tabulated differences (affected twin value minus unaffected twin value) in value of covariates in MZ and DZ twins discordant for "ever wheezed", in a co-twin case-control study I performed[13]. The affected twin listed reported a history of wheeze, while the unaffected twin denied ever wheezing.

Table 5.2 – Testicular cancer in twins versus birth order and sex of co-twin.[12]

Twin Pair	First Born	Second Born	Odds Ratio
Same Sex	39	21	0.55 (0.31–0.96)
Opposite Sex	9	22	2.50 (1.11–6.15)

There were 48-62 MZ pairs where both values were present, and 52-62 DZ pairs. Because the distribution of the skin prick test wheals is bimodal, I have used nonparametric tests to assess the within-pair differences. There are significant differences between the MZ and DZ differences for several variables.

The path model

Here is one approach to allow interpretation of the differing results seen in the different zygosity groups. It is a bivariate (actually bi-trait) "genetic correlation" path model, which assumes no gene by environment interaction.

Let X_i denote the selected trait and Y_i the covariable in relative i. The variances and covariances of X and Y in the population can be expressed as a cholesky decomposition bivariate genetic factor model containing common and unique additive genetic, shared environmental and unshared environmental components of variance.[14] I write the path coefficient from the common additive genetic component to trait Y as h_{yc}, and that from the unique additive genetic component as h_{ys}, and so on (Figure 5.1). The additive genetic factors are correlated α within the pair (where α is the coefficient of relationship), while shared environmental influences are modelled as being perfectly correlated within the pair.

Table 5.3 – Within-pair differences for asthma-discordant twins[13]

Trait	MZ Mean Difference D_{MZ}	P-value Wilcoxon test $D_{MZ} = 0$	DZ Mean Difference D_{DZ}	P-value Wilcoxon test $D_{DZ} = 0$	P-value Wilcoxon test $D_{MZ} = D_{DZ}$
Alternaria	+0.13mm	0.66	+1.17mm	0.001	0.020
Per. Ryegrass	−0.21mm	0.85	+1.98mm	0.0005	0.015
Cat hair/epi	+0.81mm	0.001	+1.57mm	0.0001	0.066
Cockroach	+0.52mm	0.009	+0.21mm	0.40	0.76
D. pter	+1.64mm	0.0002	+1.86mm	0.002	0.66
House Dust	+1.15mm	0.0001	+2.17mm	0.0001	0.24
IgE	+7 U/ml	0.016	+96 U/ml	0.003	0.42
Pack Years*	+0.53 Pkyr	0.49	+0.92 Pkyr	0.71	0.79
Height	−0.29cm	0.68	−0.8cm	0.28	0.64

* Approximate pack-years of cigarettes smoked based on duration in years and average

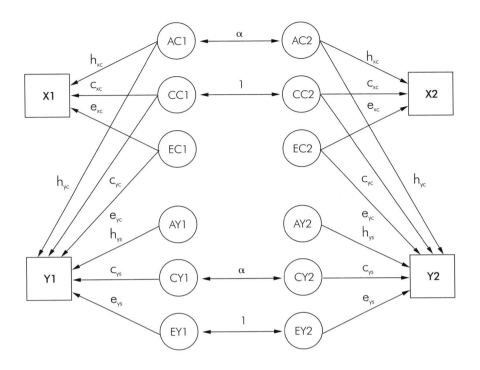

Figure 5.1 – Path diagram for Cholesky decomposition bivariate genetic factor model. α is the coefficient of relationship.

To derive the conditional expectation for the intrapair difference for trait Y (ΔY) given the pair is selected for discordance of a binary trait X (Twin 2 affected, Twin 1 unaffected), one can solve the usual regression equation:[15]

$$\boldsymbol{S}_x \mathrm{B} = \boldsymbol{S}_{xy} \tag{2}$$

where $\boldsymbol{S}_x$ is the covariance matrix for $\mathbf{X}$ $[X_1\,X_2]'$, B the regression coefficients, $\boldsymbol{S}_{xy}$ the covariances of X_1 and X_2 with ΔY (Y_2–Y_1), with X and Y expressed as deviation scores. X is coded as a dummy variable taking values of 0 and 1 for absent and present - the population mean of X is the trait prevalence (P), the variance $P(1-P)$, with an upper bound of ¼. We find:

$$\boldsymbol{S}_x = \begin{bmatrix} h_{xc}^2 + c_{xc}^2 + e_{xc}^2 \\ \alpha h_{xc}^2 + c_{xc}^2 \qquad\qquad h_{xc}^2 + c_{xc}^2 + e_{xc}^2 \end{bmatrix} \tag{3}$$

$$\boldsymbol{S}_{xy} = \begin{bmatrix} -(1-\alpha)h_{xc}h_{yc} - e_{xc}e_{yc} \\ (1-\alpha)h_{xc}h_{yc} + e_{xc}e_{yc} \end{bmatrix} \tag{4}$$

And so,

$$\Delta Y = \frac{(1-\alpha)h_{xc}h_{yc} + e_{xc}e_{yc}}{(1-\alpha)h_{xc}^2 + e_{xc}^2}(X_2-X_1) \tag{5}$$

For discordant pairs on the binary variable X, $X_2-X_1 = 1$, so for monozygotic and dizygotic (DZ) twins (as well as other first degree relatives),

$$\Delta Y_{mz} = \frac{e_{xc}e_{yc}}{e_{xc}^2} = \frac{e_{yc}}{e_{xc}}$$

$$\Delta Y_{dz} = \frac{h_{xc}h_{yc} + 2e_{xc}e_{yc}}{h_{xc}^2 + 2e_{xc}^2} \tag{6}$$

We can determine of course whether the environmental correlation is zero ($\Delta Y_{MZ} = 0$). Since the variance of ΔY is

$$\sigma_{\Delta Y}^2 = 2\frac{[(1-\alpha)h_{xc}h_{yc} + e_{xc}e_{yc}]^2}{(1-\alpha)h_{xc}^2 + e_{xc}^2} \tag{7}$$

we can solve for the genetic and environmental covariances between X and Y,

$$Cov_E = e_{xc}e_{yc} = \frac{\sigma_{\Delta YMZ}^2}{2\Delta Y_{MZ}} \tag{8}$$

$$Cov_G = h_{xc}h_{yc} = \frac{\sigma_{\Delta YDZ}^2}{\Delta Y_{DZ}} - \frac{\sigma_{\Delta YMZ}^2}{\Delta Y_{MZ}} \tag{9}$$

Interpretation of the example

I have presented only one path model for MZ and DZ cotwin-control data. There are other causational possibilities for the true underlying model, notably where one trait is an intermediary in the pathway from genes to the second trait.[14] Because of the sampling used, these can lead to different patterns of intra-pair differences in the DZ twin group. Nevertheless, one may cautiously interpret MZ-DZ differences of the type seen in the asthma study: the greater DZ than MZ mean intrapair difference seen for the first two covariates (skin tests for *Alternaria* and rye grass sensitization) is suggestive of a genetic correlation between trait and covariate. A full direction-of-causation model for that data was precluded by the ascertainment scheme.[16]

I will now describe some related models applicable when the trait and covariable are both binary/dichotomous in nature.

MODELS FOR GENE BY ENVIRONMENT AND GENE BY GENE INTERACTION

One of the applications of the twin study discussed in the report of the 1969 International Symposium "Twin Registries in the Study of Chronic Disease"[17] was the detection of gene by environment interaction. As noted earlier, this can be seen as a major motivation for the use of the cotwin-control study, in that one either wishes to control potential confounding by genes, or wishes to quantify effect modification or interaction.

In the measured genotype case, this reduces to a standard analysis, but where genotype must be inferred from history of the disease in a relative, such as a co-twin, this becomes more interesting. The symposium report suggests two designs. One is a prospective design, where MZ twin pairs are concordantly or discordantly exposed to the putative agent; the second is nominally to test for genetic corre-lation, comparing twin concordance for a secondary trait (suspected to be a *forme fruste* of the primary trait, or a pleiotropic effect of the primary trait gene) in pairs concordantly unaffected, or discordant for the primary trait.

A third approach is to utilize MZ and/or DZ disease-discordant pairs as above, either as two strata, or in a matched design incorporating an additional unrelated case (possibly from another twin pair). If the disease D can develop *only if both* a disease allele and an environmental exposure is present (both are necessary causes), then the unaffected MZ twin will always be exposure negative, while a proportion of unaffected DZ co-twins will be exposure positive.

A simple example

I will extend this situation slightly to the case of a dichotomous trait determined by a single major locus and dichotomous exposure under the following model:

The population frequency of the A allele is p, and the probability of exposure is E (exposure is uncorrelated in families). I might emphasize that the "environmental

Table 5.4 – The example model of gene by environment interaction. A diallelic trait locus has genotypes A/A, A/B and B/B with population frequencies of p^2, $2p(1-p)$ and $(1-p)^2$. In the absence of an environmental risk factor, each genotype has the same penetrance f_0, but the penetrances take different values f_2, f_1, f_0 when the risk factor is present.

Risk Factor	Gene Penetrance			Mean Risk	
	AA	AB	BB		
Exposure Positive	f_2	f_1	f_0	$p^2 f_2 + 2p(1-p)f_1 + (1-p)^2 f_0$	R_1
Exposure Negative	f_0	f_0	f_0	f_0	R_0

exposure" could in fact be a measured genotype, where epistasis is suspected. A conventional study using unrelated cases and controls will measure the mean population risk ratio R_1/R_0 (where R_0 is the mean risk in the absence of the factor, and R_1 when it is present).

A prospective twin study would examine concordance in pairs where zero, one or both members had been exposed (Table 5.5). This would detect the fact that there is no familial aggregation of risk except in pairs where both members have been exposed.

The equivalent matched case-control study is shown in Table 5.6 The estimated environmental exposure odds ratio (N_{10}/N_{01} – where N_{10} is the observed number of twin pairs where the affected twin was exposed to the risk factor and the unaffected cotwin was unexposed, and so forth) is unbiased.[18-19] Testing differences in exposure concordance $N_{11}/(N_{11}+N_{10}+N_{01})$ between two strata of pairs will allow one to test $Cov=0$, given this particular pattern of interaction, and a *large* sample.

Similar models are discussed by Khoury and James[20], Andrieu and Goldstein[21] and Gladen.[22] These authors are interested first in whether the odds ratio for the environmental exposure estimated from the table above (Table 5.6) is biased compared to that from a standard study. Khoury and James discuss an unmatched design

Table 5. 5. The expected marginal and recurrence risks for twins under the example model of gene by environment interaction.

Exposure Level of pair	Marginal Risk		Twin 2 Recurrence Risk* (Pr(T2 = Aff|T1 = Aff)
	Twin 1	Twin 2	
Both Unexposed	R_0	R_0	R_0
Twin 1 exposed	R_1	R_0	R_0
Both twins exposed	R_1	R_1	$R_1 + Cov/R_1$

*R_0 is the risk an individual expresses the trait if unexposed to the risk factor, R_1 if exposed. The binary covariance $Cov = R.2p(1-p)(p(f_2-f_1) + (1-p)(f_1-f_0))^2 + K.p^2(1-p)^2(f_2-2*f_1-f_0)^2$, R and K being the probability that the pair shares one or two alleles identical by descent and p, f_0, f_1, f_2 as per Table 5.4.

Table5. 6. – The expected proportions for a co-twin case-control study under the example model of gene by environment interaction.

Unaffected Member	Affected Member	
	Exp Positive	Exp Negative
Exposure Positive	$E^2[R_1(1-R_1)-Cov]/T$	$E(1-E)[R_0(1-R_1)]/T$
Exposure Negative	$E(1-E)[R_1(1-R_0)]/T$	$(1-E)^2[R_0(1-R_0)]/T$

* The total probability $T = R_0(1-R_0) + E(R_1-R_0)(1-2R_0)-E^2(Cov + (R_1-R_0)^2)$.
E = probability of exposure.

using a cohort of unaffected cotwins of a case suffering from a monogenic disease, but their model is useful, as it covers various patterns of G×E interaction. Gladen[22] examines the situation where both unmeasured genetic and measured environmental risk factors are correlated among family members, although independent of each other. She concluded that the effect size estimate for the environmental exposure will be biased in this case, but to a small degree in most realistic situations.

Andrieu and Goldsteins' model

These authors[21] use a simplified model for the genetic risk factor, treating it as binary ie dominant or recessive (as do Khoury and James[20]). Let

$$
\begin{aligned}
a &= \Pr(\text{Aff}\,|\,G+,E+) \\
b &= \Pr(\text{Aff}\,|\,G+,E-) \\
c &= \Pr(\text{Aff}\,|\,G-,E+) \\
d &= \Pr(\text{Aff}\,|\,G-,E-)
\end{aligned} \tag{10}
$$

where G and E are binary genetic and environmental risk factors, "+" is present, "-" is absent, with population proportions p and q, and the conditional probability a co-twin carries the disease factor G given their twin is G+, $\Pr(T1=G+\,|\,T2=G+)=r$. Then the odds ratio estimated from a co-twin case-control study will be,

$$
OR = (rz_1+z_2)/(rz_1+z_3), \tag{11}
$$

where,

$$
\begin{aligned}
z_1 &= (a-c)(d-b), \\
z_2 &= (1-d)(a+c/p-2c)+c(1-b), \text{ and} \\
z_3 &= (1-c)(b+d/p-2d)+d(1-a).
\end{aligned}
$$

When $a=c$ or $b=d$, that is when the genetic factor modulates risk only within one stratum of exposure to E (as in our example above), the matched analysis OR are an unbiased estimate of the population (mean) effect of exposure to E. With a more complex relationship, the matched analysis odds ratio can be higher or lower than population expectation.

POWER AND RELEVANCE OF THE CO-TWIN CASE-CONTROL DESIGN

For the simple G×E example we have examined, it is clear that the MZ co-twin case-control study will be more informative than a conventional study in that it enrichs the proportion of controls exposed to the susceptibility genotype. This will be particularly useful if the disease allele is rare, but needs to be balanced with the difficulty of recruiting a sufficient number of discordant twin pairs.

In the case of experimental studies, and prospective studies of quantitative traits, the paper by Christian and Kang[5] give nomograms for calculating power and relative efficiency of co-twin control versus conventional studies. For the co-twin case-control study (of dichotomous traits), there are no equivalent papers to my knowledge. Dupont[23] does presents (asymptotic) formulae for the power of the matched case-control study given the correlation within the pair for the exposure.

To perform meaningful comparisons *vis à vis* an equivalent unmatched study, one needs to specify the effect size for the risk factor of interest, the twin correlation for the risk factor, the trait heritability, the pattern of interaction, and the relative costs of ascertaining twins and cases. The relevant measure of heritability for a case-control study is the recurrence risk for the particular class of twin or population relative risk. The interaction can be summarised as being additive, multiplicative, or complex. These are probably most easily simulated for a particular study, although a generalization of the formulae presented by Andrieu and Goldstein[21] would be possible.

CONCLUSIONS

The co-twin case-control design is useful wherever a trait of interest is heritable. The results from MZ co-twin control studies are superficially easy to interpret, but may give different results from a study using unrelated individuals. Including a DZ co-twin component is common, and allows testing for some types of interaction.

In the large twin registries, it can be more powerful to ignore the family relationships and perform conventional cohort studies when data collection is cheap and complete.

If measurement is expensive however, the efficiencies of the co-twin study make it very attractive.

REFERENCES

1 Galton F. Twins, as a criterion of the relative powers of nature and nurture. Fraser's Magazine. 1876.

2 Galton F. Inquiries into human faculty and its development. London: The Eugenics Society. 1883.

3 Newman HH, Freeman FN, Holzinger KJ. Twins: a study of heredity and environment. Chicago: University of Chicago Press. 1937.

4 Gesell. Science 1942;

5 Christian JC, Kang KW. Efficiency of human monozygotic twins in studies of blood lipids. Clin Exp Metabol 1972; 21: 691–699.

6 Fisher RA. Dangers of cigarette-smoking. Brit Med J 1957; 2: 297–298.

7 Lassila R, Seyberth HW, Haapanen A, Schweer H, Koskenvuo M, Laustiola KE. Vasoactive and atherogenic effects of cigarette smoking: a study of monozygotic twins discordant for smoking. Brit Med J 1988; 297: 955–957.

8 Floderus B, Cederlof R, Friberg L. Smoking and mortality: a 21-year follow-up based on the Swedish Twin Registry. Int J Epidemiol 1988; 17: 332–340.

9 Hopper JL, Seeman E. The bone density of female twins discordant for tobacco use. New Engl J Med 1994; 330: 387–392.

10 Cicuttini FM, Baker JR, Spector TD. The association of obesity with osteoarthritis of the hand and knee in women: a twin study. J Rheumatol 1996; 23: 1221–1226.

11 Vagero D, Leon D. Ischaemic heart disease and low birth weight: a test of the fetal-origins hypothesis from the Swedish Twin Registry. Lancet 1994; 343: 260–263.

12 Swerdlow AJ, De Stavola B, Maconochie N, Siskind V. A population-based study of cancer risk in twins: relationships to birth order and sexes of the twin pair. Int J Cancer 1996; 67: 472–478.

13 Duffy DL, Mitchell CA, Martin NG. Genetic and environmental risk factors for asthma: a cotwin-control study. Am J Respir Crit Care Med 1998; 157:840-845

14 Duffy DL (1994). Biometrical genetic analysis of the cotwin control design. Behavior Genetics 24: 341–344.

15 Cliff, N. Analysing multivariate data. San Diego: Harcourt Brace Jovanovich. 1987.

16 Heath AC, Kessler RC, Neale MC, Hewitt JK, Eaves LJ, Kendler KS. Testing hypotheses about direction of causation using cross-sectional family data. Behav Genet 1993; 23: 29–50.

17 Cederlöf R, Epstein FH, Friberg LT, Hrubec Z, Radford EP. Twin registries in the study of chronic disease. Acta Med Scand Suppl 1971; 523.

18 Goldstein AM, Hodge SE, Haile RWC. Selection bias in case-control studies using relatives as the controls. Int J Epidemiol 1989; 18: 985–989.

19 Pike MC, Robins J. Re: 'Possibility of selection bias in matched case-control studies using friend controls'. Am J Epidemiol 1989; 130: 209–210.

20 Khoury MJ, James LM. Population and familial relative risks of disease associated with environmental factors in the presence of gene-environment interaction. Am J Epidemiol 1993; 137: 1241–1250.

21 Andrieu N, Goldstein AM. Use of relatives of cases as controls to identify risk factors when an interaction between environmental and genetic risk factors exists. Int J Epidemiol 1996; 25: 649–657.

22 Gladen BC. Matched-pair case-control studies when risk factors are correlated within the pairs. Int J Epidemiol 1996; 25: 420–425.

23 Dupont WD. Power calculations for matched case-control studies. Biometrics 1988; 44: 1157–1168.

6

GENERALISABILITY AND ASSUMPTIONS OF TWIN STUDIES

Kirsten Ohm Kyvik

ABSTRACT

Twin pairs are usually studied in genetic epidemiology with the purpose of estimating heritability, recurrence risks and aetiological models for diseases or traits. Being able to generalize and compare these results to those from the general population is an important part of a study. Twin studies are based on three main assumptions, some of which have been questioned.

The assumption of twinning mechanisms: The concept of dispermatic twins and twinning as a result of superfetation or superfecundation, especially if the event involves dizygotic twins with different fathers have resulted in speculation about possible genetic biases. The rarity of these events makes any bias unlikely.

The assumption of equal environment: It is assumed that monozygotic and dizygotic twin pairs share environment to the same degree and that a greater phenotypic similarity among monozygotic than dizygotic twin, results from the two-fold greater genetic similarity. If, on the other hand, monozygotic twins are treated more similarly than dizygotic twins, this would result in an overestimation of the importance of genetic effects. However, studies of IQ in twins correctly and incorrectly classified in terms of zygosity, have shown no difference between the two groups and thus no effect of parental bias.

The assumption of representativeness: The twin pregnancy itself involves some peculiarities preventing a comparison to singletons, e.g. a 3–4 week pre-term delivery, a birth weight on average 1000 grams less and a greater risk of birth complications and malformations. Thus, it is important to check that means and variances of continuous measures and for diseases, the incidence or prevalence in twins are the same as in the general population. Since there is a tendency for MZ twin pairs and female twin pairs to be more willing to volunteer it is also important to check for an effect of zygosity, and sex, in the analysis, e.g. by checking that means and variances in the groups are the same. Studies of mortality and a number of diseases, among them insulin-dependent diabetes mellitus, in the Danish Registry have shown twins to have the same prevalence of several diseases and the same mortality as the general population, rendering a comparison to the general population possible

A concern in the interpretation of twin research has been the generalisability of results from twins to the rest of the population. The scientific knowledge derived from twin studies is interesting in itself but should be understood in a greater perspective of public health. Colleagues from the clinical departments need to extrapolate this information in order to be able to offer genetic counselling to patients and to co-twins of diseased twins.

This chapter examines existing data on the generalisability of twin studies and focuses on the issues surrounding twin biology, the equal environments assumption, differences with singletons and twin-twin interactions. There is no clear evidence that deviations in the assumptions impair the ability to interpret or generalise from twin studies.

The classical twin study is based on our knowledge that twins are of two kinds. Monozygotic (MZ) twin pairs derive from a single fertilised ovum that divides and develops into two individuals, who are regarded as being genetically identical. Dizygotic (DZ) twin pairs result from the ovulation, fertilisation and implantation of two eggs, and these twins share their genes to the same extent as ordinary siblings, i.e. they have on average 50 per cent of genes in common.[1] Both types of pairs are believed to share intra-pair environment to the same degree despite the twofold difference in genetic similarity, usually referred to as the Equal Environment Assumption (EEA). The basic concepts of twin methodology are thus; (i) a greater phenotypic similarity among MZ than DZ twins must be caused by the greater genetic similarity, i.e. the trait or disease under study is influenced by genetic factors and (ii) discordance in MZ pairs can be attributed to environmental effects only.[2,3] This is the basis of what is usually called the classical twin study.

Three important issues underlie this: (i) The biology of twinning is as simple as described above; (ii) The EEA is true; (iii) Twins are a true subset of the population and representative of the target population from which the researcher has been sampling, (see also chapters 2 and 3).

BIOLOGY OF TWINNING

Monozygotic twin pairs

The most widely used explanation of biology of twinning is an important assumption because prenatal environmental exposures or genetic changes might have substantial influences on twins; which might render generalisability impossible and change our whole concept of the twin study.

The frequency of monozygotic twinning is approximately 4 per 1000 births. It is not known to be influenced by demography, heredity or maternal risk factors. The division can take place any time during the first 10–14 days after fertilisation and pending on the time of separation will result in MZ twins with separate chorion and amnions, separate amnion and common chorion or, extremely rarely, common amnion as well as chorion. The theories on the causes of twinning have included lesions of zona pellucida interfering with normal hatching, and ageing of the oocyte caused by delayed fertilisation.[4] A subsequent division of the blastocyst is thought somehow to result from this.

There are a number of rare isolated reports of genetically discordant MZ twins. A few examples: (i) different chromosome constitution caused by post-zygotic non-disjunction (even leading to different sex with one twin being a 46 XY and the other 45 X). (ii) different genomic imprinting leading to discordance for Beckwith-Wiedemann Syndrome. And (iii) skewed X chromosome inactivation leading to the differential expression of X-linked Mendelian disorders such as Duchennes muscular dystrophy.[1,4] These findings lead to the suggested hypothesis of MZ twinning being caused by a genetic change in a small group of cells, this change being recognised as alien by the blastocyst and separated from the "healthy" cells to become the co-twin.[5] This would also involve very small genetic changes as most MZ twins pairs do not demonstrate clinically obvious genetic discordance. This challenges the assumption that discordance in MZ twins can be caused by environmental differences only. If true, it would mean that heritability estimates are biased. Furthermore, it is likely that there is no single explanation as to the causes of MZ twinning. These potential problems pose challenges to twin researchers as well as molecular biologists.

Dizygotic twin pairs

DZ twinning is the result of double ovulation, fertilisation and implantation. It is known to be related to maternal age, parity and height and to genetic effects.[6] This rather straightforward biological explanation of DZ twinning has also been challenged, e.g. by the scarce evidence from animal studies of dispermatic twinning, i.e. the double fertilisation of non-independent ova, derived by the division of the primary or secondary oocytes or of the ovum immediately before fertilisation. Such DZ twins would be genetically more similar than ordinary DZ twins, i.e. the proportion of shared genes in such a pair could be expected to be between 0.5 and 0.75 instead of the average 0.5 in normal DZ twins. Although studies on rodents lend credit to the hypothesis, evidence of human dispermatic twins is still lacking.[2,4] Even if a pair of this type of twins exists in a sample of twins, the notion that this pair could have any substantial impact on the results is hard to believe.

Two other mechanisms of twinning are superfecundation, i.e. the fertilisation of simultaneously released ova by sperms released in different intercourses, and superfetation, i.e. the fertilisation of ova released in different menstrual cycles. Both of these raises the possibility of different fathers of DZ twins, i.e. a lesser genetic similarity than in ordinary DZ twins with an expected proportion of shared genes lowered to 0.25. DZ twins representing these types of twinning are hypothetical to rare, although case reports exist .[8-10] If at all present in a sample of DZ twins the effect of the dispermatic twins would most likely outbalance the effect of the other two types.[2]

The issue of chorionicity and placentation

One third of MZ pairs have totally separate amnions and chorions and their placentas may be either separate or fused, depending on the distance of implantation of the two zygotes in the uterus. Two thirds of monozygotic twin pairs have a common chorion and placenta. Of this last group approximately one percent have both a common chorion and amnion. DZ twin pairs always have separate amnions and chorions, but placentas may be fused if the implantation of the two zygotes is close (see Table 6.1).

From the twins with separate chorions about 50% have fused placentas, irrespective of zygosity. Due to this, a considerable proportion of MZ twins and 5–8 percent of DZ twins have vascular anastomoses. It has been proposed that inferences drawn from twin studies would be biased due to this adverse prenatal environment, especially in MZ twins who are slightly lighter at birth.[11]

A study of birthweight in twins with known chorionicity and placentation demonstrated no striking differences in birthweight which could be directly attributed to monochorionic or dichorionic placentation in MZ twins.[12] The distributional characteristics of birthweight in MZ pairs were quite similar to those in DZ twins with a fused placenta. Furthermore, the authors found great intra-pair variability (>500 grams) in birthweight in monochorionic, and dichorionic MZ twins with fused placenta, as well as in DZ pairs. The study demonstrates that the effects of chorionicity are subtle and not limited to the simple discussion on fetal transfusion in monochorionic twins only. The value of gathering data on chorionicity is indisputable but usually impossible. The data are not routinely entered in midwifes or doctors records and furthermore, twins are often recruited from sources, e.g. administrative registers, with no knowledge of circumstances of birth.

Table 6.1 – Zygosity and placentation. Proportions with 95% confidence intervals (adapted from [7] and [26])

Dichorionic		Monochorionic
MZ	**DZ**	**MZ**
0.31	1.00	0.69
(0.27–0.34)	(0.99–1.00)	(0.66–0.73)

THE EQUAL EVIRONMENT ASSUMPTIONS

This basic assumption of the classical twin study is also the one that has been disputed most vigorously.[13,14] Some critics of twins studies have accused twin researchers of lack of objectivity with regard to the equal environment assumption (EEA) but are themselves also rather subjective in their critique.

For many variables the assumption is valid. Both MZ and DZ twins share the womb at the same time, are exposed to environmental factors, e.g. maternal smoking or drinking, at the same time. They are raised in the same family and are of the same age. Studies have however shown that MZ twins seem to be treated more similarly by their parents than DZ twins (summarized in 3), resulting in the criticism that this leads to an overestimation of genetic effects, i.e. bias the heritability estimate upwards.

A series of studies published primarily in Behavior Genetics have tried to test the equal environment assumptions. One approach has been to study whether behavioural similarity is the same in twins with mislabelled zygosity compared to those with correct zygosity.[15,16] It was shown that the labelling had little effect on behavioural similarity and therefore supporting the assumption of equal environment. Another approach has been to test whether observed differences in environment makes a difference behaviourally.[3] One such study demonstrated that the similarity of behaviour of MZ twins was not related to how closely they shared environment. MZ twin pairs can differ in appearance, but this makes no difference to behavioural similarity ,[17,18] as might be expected if degree of resemblance determined the way parents treated MZ offspring. These studies support the validity of the equal environment assumption.

Researchers could choose to regard the equal environment as a variable to include in the analysis. Data on cohabitational history could be used to dissect whether effects attributable to common environment depends on cohabitational status in the same, or quite different ways in MZ and DZ twins.[19] A test of equality of environmental covariances in MZ and DZ twins based on an analysis of variance model, have also been suggested.[20] Including other relatives in twins studies could also be informative with regard to the effect of common environment and the equal environment assumptions.[1]

GENERALISABILITY

Twins are on average 1000 g lighter than singletons and they are born on average approximately 3 weeks pre-term. Furthermore, twin births have more frequent complications, caesarean sections and malformations than singletons. Studies on these and other obstetric and paediatric issues can thus not be generalised.

In studies on other diseases and traits the special obstetrical and paediatric issues mentioned above are not necessarily a problem, but is very important to test (i) that the disease or trait studied is the same in twins and singletons and (ii) ensure that there is no association between disease and zygosity or sex.

The first approach ensures that there is no association between the disease or trait studied and twinning itself and means that the results from the twins study can be generalised to the population. This is referred to as the generalisability of results from twin studies. The second ensures that there is no association between the disease or trait and type of zygosity or sex.

In twin studies on categorical data, where the twins can be either affected or non-affected, the prevalence or incidence of disease in twins should be compared to those in singletons. Table 6.2 illustrates this for Type 1 diabetes mellitus. No difference is found between the prevalence of this disease in young Danish MZ and DZ twins compared with the general Danish population.[21] Reliable incidence data is scarce in a lot of diseases, which limits their use. In studies on continuous data it is important to test that the mean and variances of the trait is the same in twins as in the general population. In the same study we found a mean fasting blood glucose of 10.8 mmol/l in twins with diabetes, while for singletons with the same disease the mean fasting blood glucose was 10.5mmol/l. Variances were also very similar. These findings ensure the generalisability of the findings in this study. The same procedure can be used to test for the equality of the means and variances of the different sex-by-zygosity groups and the equality of environmental covariances in MZ and DZ twins.[20]

In the Danish Twin Register[22] a number of clinically oriented studies has shown the same pattern as the example on Type 1 diabetes mellitus, i.e. most diseases have the same prevalence in twins as in the general population. Among diseases confirming these findings are Type 2 diabetes mellitus, nickel allergy and psoriasis.

Two recent studies also confirm the generalisability of results from twin studies. In the first one the authors studied mortality in Danish twins surviving the age of six years and compared it with the general population.[23] As can be seen from Figure 6.1 the mortality in MZ and DZ twins were the same as in the general population. In

Table 6.2 – Prevalence of insulin dependent diabetes mellitus in Danish singletons and twins in the age groups 0–39 years.

Singletons	MZ twins	DZ twins	Total twins*
0.0028 (0.0026–0.0030)	0.0034 (0.0023–0.0045)	0.0030 (0.0023–0.0037)	0.0033 (0.0027–0.0039)

* includes twins with unknown zygosity

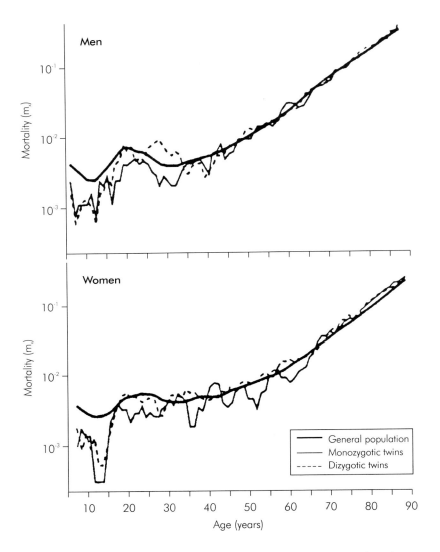

Figure 6.1 – Mortality (three year moving average) among monozygotic twins, dizygotic twins, and the general population in Denmark, age 6–89 years. Mortality was calculated for cohorts born 1870–1900.[23]

the other study[24] the fecundability, i.e. ability to conceive, measured as waiting time to pregnancy, was compared in twins and singletons. There was no decrease in the fecundability of twins compared to singletons and furthermore the distribution of waiting time to pregnancy was the same in females from MZ, DZ same-sex and DZ opposite-sex pairs.

Twin-twin interaction

The interaction between twins could theoretically influence both the inferences drawn from and the generalisability of twin studies and it is not an issue that has been widely discussed.[25] It is well-known that twins spend more time with their co-twin at the same developmental level and less time with the parents or an older child than singletons; believed to be the reason for the development of the secret language in twins. In singletons disturbed language development is often associated with socio-emotional and behavioural problems. This is not as marked in twins and most pairs outgrow their language problems. There is the possibility of (i) rivalry for the parents attention, (ii) polarisation of the two twins in a pair to strengthen individuality and (iii) dominance by one twin over the other. These last issues do also take place among ordinary siblings. Studies on socio-emotional and behavioural problems in twins and singletons have shown that twins have slightly more behavioural, but fewer emotional problems.[25] Twin-twin interaction thus is probably not a major problem interpreting the majority of twin studies.

CONCLUSION

In this chapter the assumptions of the classical twin study have been discussed, along with the issues of generalisability of twin studies and comparability between MZ and DZ twins. It seems that although all the assumptions can be questioned to some degree, so far there has been no substantial evidence that deviations from these assumptions are critical to the interpretation of twin studies.

REFERENCES

1 Martin N, Boomsma D, Machin G. A twin-pronged attack on complex traits. Nature Genetics 1997; **17**: 387–91

2 Parisi P. The twin method. In Multiple Pregnancy, Epidemiology, gestation and perinatal outcome. Keith L.G., Papiernik E, Keith D.M., Luke B (Eds). The Parthenon Publishing Group. New York, London 1995, pp 9–20

3 Plomin R, DeFries JC, McClearn GE. (1990) *Behavioral Genetics. A primer.* WH Freeman and Company, New York.

4 Bomsel-Helmreich O, Al Mufti W. The mechanism of monozygosity and double ovulation.In Multiple Pregnancy, Epidemiology, gestation and perinatal outcome. Keith L.G., Papiernik E, Keith D.M., Luke B (Eds). The Parthenon Publishing Group. New York, London 1995 pp 25–40

5 Hall J.G., Lopez-Rangel E. Embryologic development and monozygotic twinning. Acta Genet Med Gemellol 1996; **45**: 53 57

6 MacGillivray I, Samphier M, Little J. (1988) Factors affecting twinning. In *Twinning and twins* (eds. MacGillivray I, Campbell DM, Thompson B). John Wiley & Sons, Chichester, New York, Brisbane, Toronto, Singapore, 67–92.

7 Burn J, Corney G. Zygosity determination and the types of twinning. In Twinning and Twins. MacGillivray I, Campbell D.M, Thompson B (Eds). 1988. John Wiley and Sons Ltd. Pp 7–25

8 Lu HL, Wang CX, Wu FQ, Li JJ. Paternity identification in twins with different fathers. J. Forensic. Sci. 1994: 39(4): 1100–2

9 Wenk RE, Houtz T, Brooks M, Chiafari FA. How frequent is heteropaternal superfecundation? Acta Genet. Med. Gemellol, Roma 1992; 41(1): 43–7

10 Wenk RE, Houtz T, Chiafari FA, Brooks M.Superfecundation identified by HLA, protein and VNTR DNA polymorphisms. Transfus. Med. 1991; 1(4): 253–5

11 Phillips D.I.W. Twin studies in medical research: Can they tell us whether diseases are genetically determined. Lancet 1993; **341**: 1008–9

12 Corey L.A., Nance W.E., Kang K.W., Christian J.C. Effects of type of placentation on birthweight and its variability in monozygotic and dizygotic twins. Acta Genet Med Gemellol 1979; **28**: 41–50

13 Price B. Primary biases in twin studies. A review of prenatal and natal difference-producing factors in monozygotic pairs. American Journal of human genetics 1950; 2 (4): 293–352

14 Pam A, Kemker S.S, Ross C.A, Golden R. The "Equal environment assumption" in MZ-DZ twin comparisons: an untenable premise of psychiatric genetics. Acta Genet Med Gemellol 1996; **45**: 349–50

15 Scarr S. Environmental bias in twin studies. Eugen Quart 1968; 15 (1): 34–40

16 Scarr S, Carter-Saltzman L. Twin method: Defense of a critical assumption. Behavior Genetics 1979; 9 (6): 527–42

17 Matheny Jr, A.P., Wilson R.S., Dolan A.B. Relations between twins' similarity of appearance and behavioral similarity: Testing an assumption. Behavior Genetics 1976; 6 (3): 343–52

18 Plomin R., Willerman L, Loehlin J.C. Resemblance in appearance and the equal environment assumption in twin studies of personality traits. Behavior Genetics 1976: 6 (1): 43–52

19 Hopper J.L.The epidemiology of genetic epidemiology. Acta Genet Med Gemellol 1992; **41**; 261–73

20 Christian J.C. Testing twin means and estimating genetic variance. Basic methodology for the analysis of quantitative twin data. Acta Genet Med Gemellol 1979; **28**: 35–40

21 Kyvik KO, Green A, Beck-Nielsen H. (1995) Concordance rates of insulin-dependent diabetes mellitus. A population-based study of young Danish twins. *BMJ*, **311:** 913–7.

22 Kyvik KO, Christensen K, Skytthe A, Harvald B, Holm NV. (1996) The Danish Twin Register. *Dan Med Bull*, **43**: 467–70.

23 Christensen K, Vaupel J, Holm NV, Yashin AI. (1995) Twin mortality after age six: fetal origin hypothesis versus twin method. *BMJ*, **310:** 432–6.

24 Christensen K, Basso O, Kyvik KO, Juul S, Boldsen J, Vaupel JW, Olsen J. Fecundability of Female Twins. Epidemiology 1998; 9 (2): 189–192

25 Rutter M, Redshaw J. Annotation: Growing up as a twin: Twin-singleton differences in psychological development. J Child Psychol Psychiat 1991; 32 (6): 885–95

26 Kingston H.M. ABC of clinical genetics. BMJ, London 1989.

7

FETAL PROGRAMMING OR GENES?

David I. W. Phillips

ABSTRACT

Twin studies showing greater concordance in monozygous (MZ) than dizygous (DZ) twins for diseases such as non-insulin-dependent diabetes (NIDDM), hypertension and cardiovascular disease are interpreted as showing a significant genetic component in these conditions. However, recent findings that NIDDM, hypertension and circulatory disease are linked with impaired intrauterine growth suggest that these diseases may begin as a result of abnormal development during fetal life. Because genes do not appear to play a major role in controlling fetal growth, it has been suggested that these associations arise as a result of 'programming' rather than the operation of genetic factors. Programming occurs when a stimulus or insult at a critical period of development results in long term changes in physiology and metabolism. There is increasing evidence that long term alterations in the set-point of major hormonal systems may be important mediators of programming. These findings raise the possibility that MZ similarity for these conditions may, in part, be due to MZ twins sharing a more similar early environment than DZ twins. This arises because the majority of MZ twins are monochorionic, share a placenta and have vascular anastomoses between the circulation of the two twins. These anastomoses may enable hormonal influences from one twin to affect the endocrine development in the other twin and result in more similar physiological and metabolic function throughout life.

The finding of higher concordance rates in monozygous compared with dizygous twins for many of the common degenerative diseases of adult life including raised blood pressure, non-insulin-dependent diabetes (NIDDM), and cardiovascular disease have suggested a significant heritability for these conditions. However, a recent series of studies linking these conditions with growth during fetal and early neonatal life have suggested the novel hypothesis that many of these disorders may begin as a result of suboptimal development during fetal life.[1] This hypothesis originated with the observations that low birthweight or other indices of suboptimal fetal growth were associated with an increased prevalence of raised blood pressure, NIDDM and cardiovascular disease. According to the hypothesis, cardiovascular disease and its related disorders result from the persistence of fetal adaptations to an adverse early environment. These observations have important implications for twin studies as they raise the possibility of an alternative, non-genetic explanation of the high concordance in monozygous twins observed for these conditions. It is suggested that the similarity of monozygous twins could in part be due to their more similar pattern of prenatal development. This chapter will review this evidence and recent attempts to resolve the debate as to whether concordance in twins is due to shared genes or shared prenatal development.

BIRTHWEIGHT AND CARDIOVASCULAR DISEASE IN ADULT LIFE

The epidemiological studies pointing to the possible contribution of the early environment to the aetiology of cardiovascular disease and related conditions were based on studies of men and women in middle and late life whose birthweight and body measurements at birth were routinely recorded. These records were discovered by means of a systematic search of archives and record offices in Britain. One of the most complete was discovered in Hertfordshire where from 1911 onwards all babies in the county, whether born in hospital or at home, had their birthweight recorded. In a study based on 16,000 records of men and women born between 1911 and 1930 death rates fell progressively with increasing birthweight (Table 7.1).[1] The association between low birthweight and coronary artery disease has been observed in other studies in the UK and in North America.[2,3]

The association between birthweight and coronary heart disease is paralleled by similar trends in two of the major risk factors for coronary heart disease – non-insulin-dependent diabetes and hypertension.[4-6] The trends are strong. In a study of 370 men, aged 65, born in Hertfordshire the prevalence of non-insulin-dependent diabetes and impaired glucose tolerance fell from 40% among men who weighed 5.5lbs (2.54 kg) or less to 14% among those who weighed 9.5lbs (4.31 kg) or more (Table 7.2). The trends are also consistent: the relation between birth size

Table 7.1 – Death rates from coronary heart disease among 15726 men and women according to birthweight.

Birthweight in pounds (kg)	Number of people	Standardised mortality ratio	Number of deaths
≤5.5 (2.54)	765	100	57
–6.5 (2.95)	2385	81	137
–7.5 (3.41)	4947	80	298
–8.5 (3.86)	4698	74	289
–9.5 (4.31)	2056	55	103
>9.5 (4.31)	875	65	57
All	**15726**	**74**	**941**

and diabetes has been observed in a variety of populations, both European and non-European,[5] while 32 studies from around the world have demonstrated an association between low birthweight and raised blood pressure.[6] These disorders are linked with insulin resistance and it is therefore of great interest that a number of recent observations suggest that reduced fetal growth in singletons is associated with insulin resistance in the adult. The Hertfordshire Study showed that low birthweight is associated with a raised prevalence of the insulin resistance syndrome – a common disorder in adult life in which raised blood pressure, glucose intolerance and disturbed lipoprotein metabolism coincide in the same patient.[7] People with this disorder have insulin resistance and raised plasma concentrations of insulin. More recently a number of studies have specifically measured insulin action in population samples and shown that small size at birth is associated with insulin resistance in adult life.[8]

Birthweight is a summary measure of fetal growth and where more detailed anthropometric data are available they tend to show that it is disproportionate fetal growth, for example thinness or stunting at birth, rather than small size *per se* which is associated with cardiovascular risk factors and coronary artery disease in adult life.[3,9,10] Animal experiments suggest that disproportionate relationships between head size, length and weight correspond to growth failure at different periods of gestation.[11]

THE PROGRAMMING HYPOTHESIS

Because the growth of a fetus is usually limited by its nutrient and oxygen supply, these observations have led to the 'fetal origins' hypothesis of the origin of these diseases. This hypothesis proposes that adaptations made by the fetus in response to undernutrition lead to permanent changes in physiology and metabolism which in turn predispose to cardiovascular and metabolic disease in adult life. The

Table 7.2 – Prevalence of non-insulin-dependent diabetes (NIDDM) and impaired glucose tolerance (IGT) in men aged 59–70 years according to birthweight.

Birthweight in pounds (kg)	No of men	Percent with IGT or NIDDM	Odds ratio adjusted for current BMI (95% CI)
≤5.5 (2.54)	20	40	6.6 (1.5–28)
–6.5 (2.95)	47	34	4.8 (1.3–17)
–7.5 (3.41)	104	31	4.6 (1.4–16)
–8.5 (3.86)	117	22	2.6 (0.8–8.9)
–9.5 (4.31)	54	13	1.4 (0.3–5.6)
>9.5 (4.31)	28	14	1.0
All	**370**	**25**	

hypothesis is supported by animal experiments which show that fetal undernutrition programmes can lead to permanent alterations in blood pressure, cholesterol metabolism, insulin secretion and other metabolic and endocrine functions which are important in human disease.[12–14] It is also supported by a study of men and women, aged 50 years, who were born during or after the Dutch winter famine of 1944. Undernutrition at any stage of gestation was linked with reduced glucose tolerance and evidence of insulin resistance in the offspring.[15]

HOW IS DISEASE PROGRAMMED ?

Although relatively little is known, evidence from recent animal studies and preliminary evidence in humans suggests that impaired fetal nutrient supply leads to permanent alterations in the neuroendocrine development of the fetus resulting in long term changes in the set point of a number of important hormonal systems which regulate metabolism and development. Changes in the hypothalamic-pituitary adrenal axis (HPAA) appear to be particularly important. Corticosteroids are biologically potent adversely affecting carbohydrate metabolism and raising blood pressure. In animal experiments, exposure of pregnant rats to a variety of stressors including low protein diets, alcohol, or non-abortive infections lead to the birth of offspring who have increased HPAA activity and increased stress-induced corticosteroid secretion throughout adult life.[16,17] It is thought that these effects are mediated by excessive fetal corticosteroid exposure which results in persisting alterations in the activity of the HPAA. Experimental studies in rats show that these effects can be mimicked by prenatal treatment with dexamethasone. This leads to permanently increased activity of the HPAA with increased concentrations of circulating corticosteroids due to lifelong alterations in the numbers of steroid receptors in the hippocampus, which is an important site of negative feedback control. In adult life the rats become hypertensive and glucose intolerant.[18,19] Low birthweight babies have raised cortisol concentrations in umbilical cord blood and raised cortisol excretion in childhood.[20] Recently we showed that birthweight was related to adult fasting plasma cortisol concentrations in men, aged 65, in the Hertfordshire study.[21]

PROGRAMMING OR GENES?

It could be argued that the observations linking birth size with insulin resistance or diseases associated with insulin resistance may be explained by a gene or genes which are expressed as insulin resistance in both the fetus and the adult. Insulin has a central role in fetal growth ensuring that growth rates are commensurate with the nutrient supply and insulin resistance would therefore impair fetal growth as has been demonstrated in transgenic mice lacking key intermediates in insulin receptor signalling. Whilst the fetal genome undoubtedly determines growth potential in utero, the weight of evidence from animal cross breeding experiments[22], from studies of half-siblings related either through the mother or the father[23], and from embryo transfer studies[24] strongly suggests that the fetal genome plays a subordinate role in determining the growth that is actually achieved. For example, among half-siblings, related through only one parent, those with the same mother have similar birthweights, the correlation coefficient being 0.58; the birthweights of half-siblings with the same father are, however, dissimilar, the correlation coefficient being only 0.1.[23] In embryo transfer studies it is the recipient mother rather than the donor mother that more strongly influences the growth of the fetus; a fetus transferred to a larger uterus will achieve a larger birth size.[24] Thus it seems that the dominant determinant of fetal growth is the nutritional and hormonal milieu in which the fetus develops, and in particular the nutrient and oxygen supply. The importance of non-genetic effects is also supported by studies of within-pair differences in monozygous twins. In a study based on the Swedish twin register, Vagero and Leon[25] showed that among MZ twins – the shorter twin in adult life had higher coronary artery disease mortality. As the shorter twin in adult life is almost invariably the smaller twin at birth, these results suggest that the association between birthsize and coronary heart disease is not explained by genetic factors. They are further strengthened by a study of twins within the Danish twin register showing that among the 14 MZ pairs discordant for NIDDM, it was the smaller twin at birth who developed diabetes.[26]

TWIN STUDIES: GENES OR FETAL PROGRAMMING?

It is possible that the effects of the early environment and genetic influences may operate independently to determine susceptibility to cardiovascular and metabolic disease. However, the strong relationships between fetal growth and cardiovascular disease offer an alternative, non-genetic explanation of high concordance in MZ twins. It is suggested that the similarity of MZ twins may not only be due to their genetic identity but because they also share a more similar early environment than dizygous twins. How these similarities may arise is illustrated in Figure 7.1 which summarises the embryology of MZ twinning. At some stage between the zygote and the formation of the embryonic disc, the formative material divides into two parts, each giving rise to a complete embryo. If the division occurs at an early stage, the resulting MZ twins will have separate sets of fetal membranes and will in this respect resemble dizygous twins. About a third of MZ and all DZ twins are of this dichorionic type. If the division is delayed until the blastocyst has formed, the two embryos will share a chorionic membrane but will develop in two separate amniotic sacs; these twins are monochorionic and are the commonest type of monozygous twin. Occasionally the division of the formative material may be delayed until the embryonic disc has separated from the cavities that will subse-

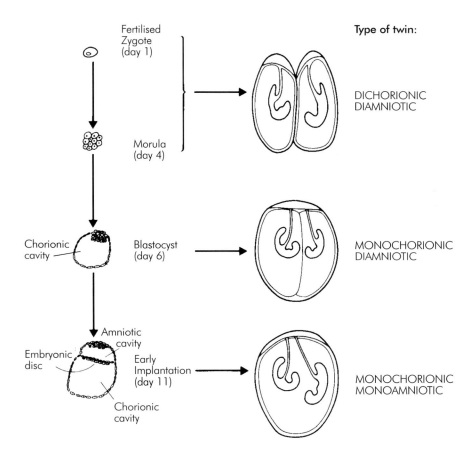

Figure 7.1 – Embryology of monozygous twinning. Division of the fertilised zygote up to the morula stage results in monozygous twins with separate chorionic and amniotic membranes which, in terms of membranes and placentation, are indistinguishable from dizygous twins. Division of the blastocyst results in monozygous twins with common chorionic but separate amniotic membranes. Rarely, division occurs after formation of the amniotic cavity, resulting in twins with shared chorionic and amniotic membranes.

quently form the amniotic and chorionic sacs. When this happens the twins will share both chorionic and amniotic sacs; few such twins survive because of the tendency of their cords to knot together. The importance of these differences lies in the effects they have on placental development. Since the placenta is formed from chorionic tissue, dizygous twins and monozygous, dichorionic twins will have separate placentas. However, monozygous, monochorionic and monoamniotic twins have common placentas. Monochorionic placentas enable the formation of vascular anastomoses between the circulations of the two twins. These anastomoses which are the rule in monochorionic placentas are usually communications joining artery to artery or vein to vein. Sometimes, however, a placental cotyledon may be supplied by an artery from one fetus and be drained by a vein to the other. In this case blood will leak from one twin to the other – the twin transfusion syndrome-resulting in the birth of a large plethoric twin and a smaller anaemic one.

DOES THE SHARED PLACENTATION OF MONOCHORIONIC TWINS MATTER?

A number of studies have shown that MZ monochorionic twins are lighter at birth than DZ or dichorionic MZ twins.[27] The differences in birthweight which amount to between 100 and 200 grams are not large but in the context of the low birth-weight of twins represent a significant reduction in birthsize. In addition, the mean duration of gestation is between 6.3 and 10.6 days shorter in monochorionic twins. Other data suggests that monochorionic twins may have differences in pregnancy outcome. Gruenwald reported a perinatal mortality rate of 7.1% in 198 mono-chorionic twins and 4.1% in 522 dichorionic twins.[28] These were confirmed by another study which reported perinatal mortality rates of 15% in monochorionic twins and 8.4% in dichorionic twins. Finally monochorionic twins appear to have a higher frequency of congenital malformations; 3.5% compared with 0.25% in dichorionic twins. These studies suggest that monochorionic twins may be more disadvantaged during gestation, perhaps as a result of competition between the twins for a limited supply of nutrients. One prediction of the fetal origins hypothesis would be that the prevalence of diabetes and cardiovascular disease would be expected to be higher in MZ than DZ twins. Two recent studies have tested this hypothesis by comparing mortality in MZ and DZ twins (see also Ch apter 5). Christensen studied 8495 Danish twins born 1870–1900. Although follow-up of the twins was low there was no difference between the mortality rates of MZ and DZ twins.[29] Likewise Vagero and Leon were unable to show differences in mortality from coronary heart disease between MZ and DZ twins in a large series of 14,786 Swedish twins born between 1886 and 1925.[25]

Because monochorionic twins have a common placenta with vascular anastomoses between the two twins, it is likely that hormones and other humoral factors will be able to diffuse easily between the circulations of the twins. Hormonal transfer from one fetus to another is a well established phenomenon in experimental animals. For example, a female fetus exposed prenatally to male hormones because it is located between two male fetuses will exhibit a variety of anatomical and behav-ioural disturbances in adult life. These effects may also occur in human pregnancies as it has been demonstrated that women who had male co-twins tended to have a more masculine psyche than did same-sex female twins.[30] If these effects occur in dichorionic, DZ twins, they are likely to be more marked in monochorionic twins and may affect a wide variety of hormonal axes. As there is increasing evidence that the programming of endocrine axes is an important mechanism linking the early environment with adult cardiovascular disease, it is possible that the similar disease susceptibility of pairs of monochorionic twins could, at least in part, result from prenatal linking of the development of hormonal axes as a result of the vascular connections between the twins. As most adult twin studies do not have data on the type of placentation, the extent of these effects is not known. However, if the disease concordance in monochorionic, MZ twins could be shown to be different from concordance in dichorionic MZ twins, it would suggest the operation of these non-genetic effects. It is, therefore, of great interest that data from the National Heart Lung and Blood Institute twin study, have suggested that the type of placentation has an important influence on the levels of serum cholesterol. These studies were based on the initial observation that the variance of cord blood cholesterol measurements in dichorionic, MZ twins was more than five times the variance in monochorionic twins. Similar findings were implied from a study of adult twins.[31] Although there was no information on the placental type in these twins, this was inferred from examination of the twins' finger and palm prints which previous studies had shown were related to the type of placentation (see Chapter 9).

CONCLUSIONS

Increasing evidence points towards the early environment as an important determinant of susceptibility to adult cardiovascular and metabolic disease. This data showing links between birthsize and the prevalence of many of these conditions is not easily explained by genetic factors. In addition, recent evidence suggests that the programming of major hormonal systems during fetal development is a plausible mechanism linking events in early life with these adult outcomes. These findings raise the possibility that MZ similarity for these conditions may in part be due to MZ twins having a more similar early environment than DZ twins. It is suggested that this arises because the majority of monozygous twins, being monochorionic, share a placenta and have vascular anastomoses between the circulations of the two twins. These anastomoses may enable hormonal influences from one twin to affect the development of endocrine axes in the other twin and result in more similar physiological and metabolic function, and hence disease susceptibility, throughout life.

REFERENCES

1 Barker DJP. Fetal origins of coronary heart disease. *Br Med J* 1995;**311**: 171–174.

2 Rich-Edwards JW, Stampfer MJ, Manson JE, et al. Birth weight and risk of cardiovascular disease in a cohort of women followed up since 1976. *Br Med J* 1997;**315**: 396–400.

3 Barker DJP, Osmond C, Simmonds SJ, Wield GA. The relation of small head circumference and thinness at birth to death from cardiovascular disease in adult life. *Br Med J* 1993;**306**: 422–426.

4 Barker DJP, Bull AR, Osmond C, Simmonds SJ. Fetal and placental size and risk of hypertension in adult life. *Br Med J* 1990;**301**: 259–262.

5 Phillips DIW. Insulin resistance as a programmed response to fetal undernutrition. *Diabetologia* 1996;**39**: 1119–1122.

6 Law CM, Sheill AW. Is blood pressure inversely related to birthweight? The strength of evidence from a systematic review of the literature. *J Hypertens* 1996;**14**: 935–941.

7 Barker DJP, Hales CN, Fall CHD, Osmond C, Phipps K, Clark PMS. Type 2 (non-insulin dependent) diabetes mellitus, hypertension and hyperlipidemia (syndrome X): relation to reduced fetal growth. *Diabetologia* 1993;**36**: 62–67.

8 Phillips DIW, Barker DJP, Hales CN, Hirst S, Osmond C. Thinness at birth and insulin resistance in adult life. *Diabetologia* 1994;**37**: 150–154.

9 Barker DJP, Godfrey KM, Osmond C, Bull A. The relation of fetal length, ponderal index, and head circumference to blood pressure and the risk of hypertension in adult life. *Paed Perinat Epidemiol* 1992;**6**: 35–44.

10 Phipps K, Barker DJP, Hales CN, Fall CHD, Osmond C, Clark PMS. Fetal growth and impaired glucose tolerance in men and women. *Diabetologia* 1993;**36**: 225–228.

11 Owens JA, Owens PC, Robinson JS. Experimental fetal growth retardation: metabolic and endocrine aspects. In: Gluckman PD, Johnston BM, Nathanielsz PW, eds. *Advances in fetal physiology*. Ithaca, New York : Perinatology Press 1989;263–286.

12 Langley SC, Jackson AA. Increased systolic blood pressure in adult rats induced by fetal exposure to maternal low protein diets. *Clin Science* 1994;**86**: 217–222.

13 Langley SC, Browne RF, Jackson AA. Altered glucose tolerance in rats exposed to maternal low protein diets in utero. *Comp Biochem Physiol* 1994;**109A**: 223–229.

14 Mott GE, Lewis DS, McGill HCJ. Deferred Effects of Preweaning Nutrition on Lipid Metabolism. *Annals NY Academy Sci* 1991;**623**: 70–80.

15 Ravelli ACJ, van der Meulen JHP, Michels RPJ, et al. Glucose tolerance in adults after prenatal exposure to famine. *Lancet* 1998;**351**: 173–177.

16 Barbazanges A, Piazza PV, Le Moal M, Maccari S. Maternal glucocorticoid secretion mediates long-term effects of prenatal stress. *J Neurosci* 1996;**16**: 3943–3949.

17 Meaney MJ, Viau V, Bhatnagar S, et al. Cellular mechanisms underlying the development and expression of individual differences in the hypothalamic-pituitary-adrenal stress response. *J Steroid Biochem Molec Biol* 1991;**39**: 265–274.

18 Levitt NS, Lindsay RS, Holmes GE, Seckl JR. Dexamethasone in the last week of pregnancy attenuates hippocampal glucocorticoid receptor gene expression and elevates blood pressure in the adult offspring of rats. *Neuroendocrinol* 1996;**64**: 412–418.

19 Lindsay RS, Lindsay RM, Waddell B, Seckl JR. Programming of glucose tolerance in the rat; role of placental 11b-hydroxysteroid dehydrogenase. *Diabetologia* 1996;**39**: 1299–1305.

20 Clark PM, Hindmarsh PC, Sheill AW, Law CM, Honour JW, Barker DJP. Size at birth and adrenocortical function in childhood. *Clin Endocrinol* 1996;**45**: 721–726.

21 Phillips DIW, Barker DJP, Fall CHD, Seckl JR, Whorwood CB, Wood PJ, Walker BR. Elevated plasma cortisol concentrations: a link between low birth-weight and the insulin resistance syndrome. *J Clin Endocrinol Metab* 1998;**83**: 757–760.

22 Ounsted M, Scott A, Ounsted C. Transmission through the female line of a mechanism constraining fetal growth. *Ann Human Biol* 1986;**13**: 143–151.

23 Morton NE. The inheritance of human birth weight. *Ann Hum Genet* 1955;**19**: 262–268.

24 Snow MHL. Effects of genome on fetal size at birth. In: Sharp F, Fraser RB, Milner RDG, eds. *Fetal growth. Proc 20th Study Group, RCOG.* London: Royal College of Obstetricians and Gynaecologists, 1989;1–11.

25 Vagero D, Leon D. Ischaemic heart disease and low birth weight: a test of the fetal-origins hypothesis from the Swedish twin registry. *Lancet* 1994;**343**: 260–263.

26 Poulsen P, Vaag AA, Kyvic KO, Moller-Jensen D, Beck-Nielsen H. Low birth-weight is associated with NIDDM in discordant monozygotic and dizygotic twin pairs. *Diabetologia* 1997;**40**: 439–446.

27 Phillips DIW. Twin studies in medical research: can they tell us whether diseases are genetically determined? *Lancet* 1993; **341**: 1008–1009.

28 Gruenwald P. Environmental influences on twins apparent at birth. *Biol Neonate* 1970;**15**: 79–93.

29 Christensen K, Vaupel JW, Holm NV, Yashin AI. Mortality among twins after age 6: fetal origins hypothesis versus twin method. *Br Med J* 1995;**310**: 432–436.

30 Miller EM, Martin N. Analysis of the effect of hormones on opposite-sex twin attitudes. *Acta Genet Med Gemellol* 1995;**44**: 41–52.

31 Reed T, Christian JC, Wood PD, Schaefer EJ. Influence of placentation on high density lipoproteins in adult males: The NHLBI twin study. *Acta Genet Med Gemellol* 1991;**40**: 353–359.

8

TWIN AND SIB-PAIR STUDIES IN DEVELOPING COUNTRIES

Xiping Xu
Changzhong Chen
Tianhua Niu
Binyan Wang
John Rogus
Nicholas J. Schork
Anyun Wang

ABSTRACT

The rapid advances of molecular genetic technologies have made it increasingly feasible to isolate and characterize genes influencing complex diseases. However, the classic linkage approach, which has been successful in identifying genetic causes of Mendelian disorders such as cystic fibrosis and Huntington's disease, has its limitations when applied to the dissection of complex traits such as essential hypertension and asthma. Appropriate study designs are essential to unravel molecular mechanisms underlying these disorders in order to overcome many difficulties such as the polygenic nature of inheritance, locus heterogeneity, and gene-environment interactions. Developing countries offer unique opportunities in this regard. To take China as an example, these studies can be effectively performed for several reasons: (1) the vast population size (about 1.3 billion people) has made twin and sib-pair studies feasible for many common diseases; (2) relative genetic homogeneity has been preserved in remote regions; (3) population stratification is distinct due to social/geographic diversity; (4) remarkable urban/rural and geographic contrasts in terms of environmental exposures and disease occurrence exist; (5) family members tend to stay congregated; (6) family-based genetic studies are cost-effective; and (7) there is less drug intervention. Developing countries such as China may provide novel insights into the pathophysiology of common disorders. Ethical, cultural and political differences and intellectual right property sharing should be considered in any collaboration with a developing country.

Systematic applications of modern molecular genetic tools have accelerated the process of identifying genes responsible for single gene disorders, such as cystic fibrosis or Huntington's disease. However, unlike monogenic diseases, there are difficulties in elucidating the genetic basis of common disorders. The classic parametric linkage approach, which is suitable for mapping genes underlying Mendelian traits, is problematic when applied to the genetic dissection of complex traits. The factors that contribute to the difficulty in identifying genes for common disorders are multiple, and include the following: (1) their nature as quantitative traits; (2) their polygenic etiology; (3) their genetic or locus heterogeneity; (4) multifactorial (ecogenetic) interactions between genetic and environmental factors; (5) gene-gene interactions (epistasis); (6) incomplete genetic penetrance or delayed age of onset; and (7) high frequency of disease-associated alleles.[1,2]

Twin and sib pair study designs are appropriate in complex disease gene mapping and developing countries can often provide family resources that are particularly well-suited to these studies. At the Program for Population Genetics (PPG) of the Harvard School of Public Health, collaborative research in developing countries such as China has been performed on disorders including asthma, essential hypertension, osteoporosis, and type II diabetes.

TWIN STUDIES

The first example of a twin study in developing countries is Chatterjee and Das's study of the relative genetic and environmental contributions in lung function variability in 54 twin pairs from India.[3] Thirty pairs of MZ twins and 24 pairs of DZ twins were examined. Within-pair variances for a variety of lung function measurements including vital capacity (VC), forced vital capacity (FVC), forced expiratory volume in 1 second (FEV 1), and peak expiratory flow rate (PEFR) were significantly smaller ($p < 0.01$) in MZ twins than in DZ twins. Within-pair correlations were all greater in MZ than DZ twins, and high levels of heritability (23–99%) were shown for most of the measurements. Residual values adjusted for physical covariates showed similar results to unadjusted values in most cases. These data indicate that major lung function measurements are possibly influenced more by genetic than environmental factors.

As the second example, in Taipei City, Taiwan, Yu et al.[4] studied a total of 339 same-sexed Chinese twin neonates [274 MZ and 65 DZ twin pairs] born in four major general teaching hospitals. Both systolic blood pressure (SBP) and diastolic blood pressure (DBP) were measured by Doppler blood pressure monitor. Within-pair mean squares of SBP and DBP were consistently smaller in MZ than DZ twins at ages one month and over. The findings remained unchanged after the adjustment for the effects of age, sex, gestational age, placentation and physical state during blood pressure measurement. The heritability estimates for adjusted SBP and DBP ranged from 0.29 to 0.55 and from 0.27 to 0.45, respectively, for infants at ages two months and over. The study indicates an important genetic component of blood pressure during infancy.

More recently, a total of 129 twin pairs were studied from Wangjiang county and Yuexi county in Anhui province, China by the PPG in collaboration with Anhui Medical University. After obtaining informed consent from all individuals enrolled, a questionnaire was administered, and a screening exam was conducted as described previously.[5,6] To identify the MZ and DZ twins, we performed genotype analysis using 8 highly polymorphic microsatellite markers from the Weber screening set 9.0:[7] A total of 52 MZ twin pairs were identified, showing the same genotypes between each other for all the 8 markers. Table 8.1 compared the unadjusted heritability estimates obtained from our investigation and those of previous U.S. studies. A strong genetic influence in all the quantitative traits examined was clearly shown (Table 8.1), and further molecular genetic analyses of these continuous phenotypes should be explored.

SIB PAIR STUDIES

Sib pair studies have been conducted in many countries (both developed and developing) around the world. One group of these studies has investigated the role of the angiotensinogen gene (AGT) in essential hypertension. Previous investigations internationally (St. Vincent and the Grenadines, Taiwan, and European countries) have revealed conflicting results. Among African Caribbeans from St. Vincent and the Grenadines, Caulfield and colleagues[8] tested for linkage by genotyping the AGT 3' GT-repeat marker in 63 affected sibling pairs with hypertension. They found significant support for linkage, suggesting that AGT may play a major role in essential hypertension in African Caribbeans. In another

Table 8.1 – The comparison of heritability estimates (based on the difference of the intraclass correlation coefficients of the MZ and DZ twin pairs) with respect to quantitative phenotypes between studies in China and in the United States.

Trait	China					US				
	MZ		DZ		h2	MZ		DZ		h2
	Male	Female	Male	Female		Male	Female	Male	Female	
Height	62	40	87	65	0.82	3948	0	4194	0	0.80[33]
Weight	62	40	87	65	0.59	3948	0	4194	0	0.81[33]
BMI	62	40	87	65	0.49	3948	0	4194	0	0.84[33]
Heart Rate	62	40	87	65	0.70	40	40	60	60	0.65[34]
Forearm proximal BMD	62	40	87	65	0.52	0	172	0	78	0.40[35]
Forearm distal BMD	62	40	87	65	0.57	0	172	0	78	0.74[35]
FEV1	26	24	36	33	0.66	254	0	282	0	0.91[36]
FVC	26	24	36	33	0.75	254	0	282	0	0.77[36]
HDL-C	62	40	87	65	0.70	146	0	162	0	0.74[37]

Note: Although heritability estimates are relatively similar in the Chinese and US studies it has to be noted that heritabilities are not absolute properties of (physical) characteristics. They are functions of the impact of genetic and environmental factors on a trait for a given population in particular circumstances. In other words heritabilities are population specific. This implies that population-specific genetic or environmental factors may play a role in determining the traits of interest.

study carried out in Taiwan, Wu and colleagues[9] found no evidence that AGT plays a role in blood pressure variation among 222 individuals (126 nondiabetic offspring and 96 parents) in 48 Taiwanese families with non-insulin dependent diabetes mellitus (NIDDM). Similarly, in a collaborative study among nine European centers, Brand et al.[10] performed linkage analysis on the largest panel of hypertensive sib pairs to date (n=630). Their results were also negative both in the whole panel of sibships and in subgroups of families stratified by certain previously defined criteria.

Recently, we (PPG in collaboration with Anhui Medical University) tested AGT using 310 concordant affected sibling pairs from Anqing, China. One of our goals was to test for excess allele sharing among hypertensive siblings genotyped using two GT-repeat microsatellite markers (AGT-GT and HLX1-GT) and also two diallelic markers (T174M and M235T) for AGT.[11] The findings revealed no evidence for linkage of the AGT markers to essential hypertension in the population studied.[11]

Besides AGT, previous reports indicate that genes encoding renin, the β- and γ-subunits of the epithelial sodium channel (β,γ ENaC), α-adducin, and kallikrein (KLK) may be linked to essential hypertension in several human populations.[12–16] To test these findings in Chinese, we performed affected sib pair analysis on microsatellite markers in the immediate vicinity of the genes encoding renin, β and γ ENaC, α-adducin, and KLK.[17] The results revealed no evidence for linkage of any of these genes to primary hypertension in the population studied.

Taken together, results of our studies using the affected sib pair approach, provide no support for a material contribution of the AGT, renin, β, γ ENaC, α-adducin, or KLK genes to the pathogenesis of essential hypertension among Chinese. These findings are compatible with the notion of ethnic heterogeneity of disease-relevant genes that play a role in the causation of common disorders, which will be discussed below.

ADVANTAGES OF GENETIC STUDIES IN DEVELOPING COUNTRIES

Finding Population-Specific Genes

Finding population-specific genes is one of the essential goals for human genetics. Racial/ethnic differences are likely to exist with regard to the distribution/prevalence of individual genetic factors contributing to a complex disease. For example, the allele frequency of the M235T variant of the angiotensinogen gene was much higher in Chinese[11] (79%), Japanese[18] (75%–89%), and African Americans[19] (82%–83%), than in Caucasians[20] (38%–47%). As a result, findings of association studies conducted in different ethnic populations differ from one to another.[18–20] Recently, the G460W variant of the α-adducin gene, which encodes a membrane skeleton protein involved in ion transport, was associated with essential hypertension in a Caucasian population.[21] However, Kato et al.[22] showed that the W460 variant appeared to be relatively more prevalent in the Japanese (54% – 60%) compared to Caucasians (13%-23%). Moreover, Kato and colleagues did not find a significant association between the G460W polymorphism and essential hypertension in Japanese.[22] Thus, ethnic variations might account for the discrepancy of the results conducted in different populations from different countries. However, this is not to say that genetic causes of common diseases are mutually exclusive in different populations. For example, the p53 gene encodes a tumor suppressor that plays a pivotal role in cell cycle control and p53 mutations at codon 249 have been shown to cause human hepatocellular carcinoma in both Chinese and South African populations.[23,24]

Generalizablility of Known Genes to Other Racial/Ethnic Populations

Generalizablility of known genes to other racial/ethnic populations varies depending on the contribution of the gene to the complex trait. For genes that have small or moderate effects on the disease manifestation, conclusions based on the investigations in one ethnic group or population will not necessarily reflect the presumably more diverse mix of disease-relevant gene variants in other populations of different ethnicity. However, for genes of substantial effects, many studies have shown that finding an appropriate country to study a disease (e.g., Huntington's disease in Venezuela,[25,26] insulin resistance in Taiwan[27,28]) can accelerate the identification of susceptibility genes for diseases relevant to other races. Furthermore, even if the susceptible genes isolated from populations of developing countries differ from those of the developed countries, such discoveries could still shed new light on the biological mechanisms of complex disorders. Conversely, after candidate variants have been identified, their clinical values for prediction of common diseases could be tested in the developed countries.

Other Advantages

There are some other advantages of conducting genetic studies in developing countries. To give our experience in China as an example, these include: (1) Populations in developing countries (e.g. rural populations in China) have remained relatively stable and homogeneous, and population isolates may exist. (2) Family members tend to stay congregated and are often very cooperative and willing to participate in genetic studies. (3) Large-scale genetic epidemiological study and phenotype data collection are cost-effective due to low labor costs. (4) A hierarchical three-tier (county, township, and village) medical service system, which serves as the backbone of health care delivery in rural areas, has been established for 25 years in China. (5) Population stratification is distinct due to social/geographic diversity. (6) Striking urban/rural and geographic contrasts in terms of environmental exposures and disease occurrence exist. (7) There is less drug intervention in developing countries such as China compared to developed countries.

Issues of genetic studies in developing countries

Political issues

Support from government officials, local hospitals, and residents, is important for a successful completion of genetic studies in developing countries. Political stability and the restrictions of national policies/regulations in developing countries are critical factors to be taken into account in planning a large-scale population genetic project. In addition, researchers should have a full understanding of governments' long-term health policy and objectives.

Cultural and Ethical Issues

Studies in developing countries are often associated with tremendous cultural challenges including low education level of research populations, existence of multiple dialects and geographic distance, and limited financial and technical support. Population geneticists can enhance their chances of success by defining priorities, choosing critical and feasible projects, taking advantage of a country's unique features, and paying attention to quality control. Research training and education are also integral components, which may consist of: [1] lectures/work-shops; [2] scholar/student exchanges; and [3] adjunct faculty appointments. In addition, access to updates of literature and the use of internet may also help to build up research capacity for developing countries and consequently to improve the quality of their research.

Ethical questions are also important issues for population-based genetic studies in developing countries. As a rule, one goal of every investigation should be to improve the health of the study subjects and to help prevent the occurrence of common diseases in the communities under investigation. Individual autonomy should be protected through appropriate informed consent process, and the confidentiality of research information must be preserved. Ethical concerns are especially critical for genetic studies of behavioral disorders and psychiatric diseases and, in all cases, an individual's privacy regarding genetic information needs to be well safeguarded. In addition, intellectual right property should be shared by researchers from both the developed and the developing countries in international collaborations.

CONCLUSIONS

Recent advances of genetic epidemiologic investigation into genes contributing to inherited complex diseases have offered the promise of improved disease diagnosis, prevention, and treatment through genetic testing and mechanism-based therapy.[29] Because of the very nature of complex traits, mapping a relevant gene to a certain interval usually require large sample sizes and a dense genetic linkage map.[30–32] Huge population size in developing countries such as China provide a special opportunity for achieving such large sample sizes of sib pairs. From our experience, it is more cost-effective to collect the intermeidate phenotype data from populations selected from developing countries. Furthermore, it is of great importance to reproduce a significant linkage in a separate population, and collaboration with developing countries will enable researchers to replicate linkage results in an independent population. This type of collaborative research can also offer many benefits to the host countries: high-level technical training, learning opportunities and access to top-notch health care. Thus, when conducted properly, twin and sib-pair studies in developing countries will not only expedite the efforts of genetic dissection of complex traits, but also provide benefits to the residents of the developing countries.

ACKNOWLEDGMENTS

We gratefully acknowledge the assistance and cooperation of the faculty and staff of the Anhui Medical University, Anqing Public Health Bureau and local hospitals. We are thankful to all the participants in our study. This work is supported in part by NIH Grant RO1 HL/AI 56371–01A1.

REFERENCES

1 Lander ES, Schork NJ. Genetic dissection of complex traits. *Science* 1994;265: 2037–2048.

2 Schork NJ. Genetics of complex disease: approaches, problems, and solutions. Am J Respir Crit Care Med 1997; 156: 5103-9

3 Chatterjee S, Das N. Lung function in Indian twin children: comparison of genetic versus environmental influence. *Ann Hum Biol 1995*;22: 289–303.

4 Yu MW, Chen CJ, Wang CJ, Tong SL, Tien M, Lee TY, Lue HC, Huang FY, Lan CC, Yang KH, et al. Chronological changes in genetic variance and heritability of systolic and diastolic blood pressure among Chinese twin neonates. *Acta Genet Med Gemellol (Roma)* 1990;39: 99–108.

5 Xu X, Niu T, Christiani DC, Weiss ST, Zhou Y, Chen C, Yang J, Fang Z, Jiang Z, Liang W, Zhang F. 1996. Environmental and occupational determinants of blood pressure in rural communities in China. *Annals of Epidemiol* 1997;7: 95–106.

6 Niu T, Xu X, Rogus J, Zhou Y, Chen C, Yang J, Fang Z, Schmitz C, Zhao J, Rao VS, Lindpaintner K. Angiotensinogen gene and hypertension in Chinese. *J Clin Invest* 1998;101: 188–194.

7 Yuan B, Vaske D, Weber JL, Beck J, Sheffield VC. Improved set of short-tandem-repeat polymorphisms for screening the human genome. *Am J Hum Genet* 1997;60: 459–460.

8 Caulfield M, Lavender P, Newell-Price J, Farrall M, Kamdar S, Daniel H, Lawson M, De Freitas P, Fogarty P, Clark AJ. Linkage of the angiotensinogen gene locus to human essential hypertension in African Caribbeans. *J Clin Invest* 1995;96: 687–692.

9 Wu DA, Bu X, Warden CH, Shen DD, Jeng CY, Sheu WH, Fuh MM, Katsuya T, Dzau VJ, Reaven GM, Lusis AJ, Rotter JI, Chen YD. Quantitative trait locus mapping of human blood pressure to a genetic region at or near the lipoprotein lipase gene locus on chromosome 8p22. *J Clin Invest* 1996;97: 2111–2118.

10 Brand E, Chatelain N, Keavney , Caulfield M, Citterio L, Connell J, Grobbee D, Schmidt S, Schunkert H, Schuster H, Sharma AM, Soubrier F. Evaluation of the angiotensinogen locus in human essential hypertension: a European study. *Hypertension* 1998;31: 725–729.

11 Niu T, Xu X, Rogus J, Zhou Y, Chen C, Yang J, Fang Z, Schmitz C, Zhao J, Rao VS, Lindpaintner K. Angiotensinogen gene and hypertension in Chinese. *J Clin Invest* 1998;101: 188–194.

12 Barley J, Carter ND, Cruickshank JK, et al. Renin and atrial natriuretic peptide restriction fragment length polymorphisms: association with ethnicity and blood pressure. *J Hypertens* 1991;9: 993–996.

13 Shimkets RA, Warnock DG, Bositis CM, Nelson-Williams C, Hansson JH, Schambelan M, Gill JR, Ulick S, Milora RV, Findling IW et al. Liddle's syndrome: heritable human hypertension caused by mutations in the β subunit of the epithelial sodium channel. *Cell* 1994;79: 407–414.

14 Hansson JH, Nelson-Williams C, Suzuki H, Schild L, Shimkets R, Lu Y, Canessa C, Iwasaki T, Rossier , and RP Lifton. Hypertension caused by a truncated epithelial sodium channel S0001TgS0000T: genetic heterogeneity of Liddle's syndrome. *Nat Genet* 1995;11: 76–82.

15 Berry TD, Hasstedt SJ, Hunt SC, Wu LL, Smith J, Ash KO, Kuida H, Williams RR. A gene for high urinary kallikrein may protect against hypertension in Utah kindreds. *Hypertension* 1989;13: 3–8.

16 Casari G, Barlassina C, Cusi D, Zagato L, Muirhead R, Righetti M, Nembri P, Amar K, Gatti M, Macciardi F, et al. Association of the alpha-adducin locus with essential hypertension. *Hypertension* 1995;25: 320–326.

17 Niu T, Xu X, Cordell HJ, Rogus J, Rao VS, Zhou Y, Fang Z, Lindpaintner K. Linkage analysis of candidate genes and gene-gene interactions in Chinese hypertensive si pairs. *Hypertension* (tentatively accepted).

18 Hata, A., Namikawa, C., Sasaki, M., Sato, K., Nakamura, T., Tamura, K., and J.-M. Lalouel. Angiotensinogen as a risk factor for essential hypertension in Japan. *J Clin Invest* 1994;93: 1285–1287.

19 Rotimi C, Morrison L, Cooper R, Oyejide,C, Effiong E, Ladipo M, Osotemihen BO, Ward R. Angiotensinogen gene in human hypertension: lack of an association of the 235T allele among African Americans. *Hypertension* 1994;24: 591–594.

20 Jeunemaitre X, Inoue I, Williams C, Charru A, Tichet J, Powers M, Sharma AM, Gimenez-Roqueplo A-P, Hata A, Corvol P, Lalouel J-M. Haplotypes of angiotensinogen in essential hypertension. *Am J Hum Genet* 1997;60: 1448–1460.

21 Cusi D, Barlassina C, Azzani T, Casari G, Citterio L, Devoto M, Glorioso N, Lanzani C, Manunta P, Righetti M, Rivera R, Stella P, Troffa C, Zagato L, Bianchi G. Polymorphisms of alpha-adducin and salt sensitivity in patients with essential hypertension. *Lancet* 1997;349: 1353–1357.

22 Kato N, Sugiyama T, Nabika T, Morita H, Kurihara H, Yazaki Y, Yamori Y. Lack of association between the alpha-adducin locus and essential hypertension in the Japanese population. *Hypertension* 1998;31: 730–733.

23 Hsu IC, Metcalf RA, Sun T, Welsh JA, Wang NJ, Harris CC. Mutational hotspot in the p53 gene in human hepatocellular carcinomas. *Nature* 1991;350: 427–428.

24 Bressac B, Kew M, Wands J, Ozturk M. Selective G to T mutations of p53 gene in hepatocellular carcinoma from southern Africa. Nature 1991;350: 429–431.

25 Kremer B, Weber B, Hayden MR. New insights into the clinical features, pathogenesis and molecular genetics of Huntington disease. *Brain Pathol* 1992;2: 321–35.

26 Penney J Jr, Young A, Shoulson I, Starosta-Rubenstein S, Snodgrass SR, Sanchez-Ramos J, Ramos-Arroyo M, Gomez F, Penchaszadeh G, Alvir J, et al. Huntington's disease in Venezuela: 7 years of follow-up on symptomatic and asymptomatic individuals. *Mov Disord* 1990;5: 93–9.

27 Wu HP, Tai TY, Chuang LM, Chiu KC, Lin BJ. CA-repeated microsatellite polymorphism of the glucokinase gene and its association with non-insulin-dependent diabetes mellitus in Taiwanese. *Diabetes Res Clin Pract* 1995;30: 21–6.

28 Chuang LM, Wu HP, Chiu KC, Lai CS, Tai TY, Lin BJ. Mitochondrial gene mutations in familial non-insulin-dependent diabetes mellitus in Taiwan. *Clin Genet* 1995;48: 251–4.

29 Struk , Cai L, Niu T, Rubattu S, Kreutz R, Lee M-A, Lindpaintner K. Genetic linkage analysis in hypertension: principles and practice. In: Lindpaintner K, Ganten D (eds): Molecular Biology and Genetics of Hypertension. London: Current Science Publishers, 1996: pp8–11.

30 Weissenbach J, Gyapay G, Di C, Vignal A, Morissette J, Millasseau P, Vaysseix G, Lathrop M. A second-generation linkage map of the human genome. *Nature* 1992;359: 794–801.

31 Hastbacka J, de la Chapelle A, Kaitila I, Sistonen P, Weaver A, Lander E. Linkage disequilibrium mapping in isolated founder populations: diastrophic dysplasia in Finland. *Nat Genet* 1992;2: 204–211.

32 Kruglyak L, Lander ES. High-resolution genetic mapping of complex traits. *Am J Hum Genet* 1995;56: 1212–1223.

33 Stunkard AJ, Foch TT, Hrubec Z. A twin study of human obesity. *JAMA* 1986;256: 51–54. 34.

34 Ditto B. Familial influences on heart rate, blood pressure, and self-report anxiety responses to stress: results from 100 twin pairs. *Psychophysiology* 1993;30: 635–645.

35 Hustmyer FG, Peacock M, Hui S, Johnston CC, Christian J. Bone mineral density in relation to polymorphism at the vitamin D receptor gene locus. *J Clin Invest* 1994;94: 2130–2134.

36 Hubert H, Fabsitz RR, Feinleib M, Gwinn C. Genetic and environmental influences on pulmonary function in adult twins. *Am Rev Respir Dis* 1982;125: 409–415.

37 Hunt SC, Hasstedt SJ, Kuida H, Stults BM, Hopkins PN, Williams RR. Genetic heritability and common environmental components of resting and stressed blood pressures, lipids, and body mass index in Utah pedigrees and twins. *Am J Epidemiol* 1989;129: 625–38.

9

COMPARISON OF ANALYSIS OF VARIANCE AND LIKELIHOOD MODELS OF TWIN DATA ANALYSIS

Joe C. Christian
Christopher J. Williams

ABSTRACT

Monozygotic (MZ) and dizygotic (DZ) twins are commonly used to estimate sources of genetic or environmental variation in cotwin control studies or to estimate population variance parameters by analysis of variance (ANOVA) or likelihood models. Both methods need simplifying assumptions to yield unbiased estimates and both use preliminary tests to assess the validity of assumed models. Population random mating and selection, no gene-environmental interactions and equal zygosity environmental covariances are usually assumed. It is possible to test for departures from normality, equality of zygosity means and equality of zygosity variances. In simulation power comparisons of the two methods, ANOVA was observed to have a slight advantage in testing for covariance and the likelihood method in testing for additive genetic variance. Both methods have relatively little power to detect gene interactions, and several data sets have provided evidence for either gene interactions or unequal zygosity common environmental variance. Prenatal environmental influences, as measured by placental type or dermatoglyphic variation, persist later in life and have been shown to cause unequal total and common environmental variances for the two zygosities. Studies of EEG power, evoked potentials and eye movements have suggested the presence of either non-additive genetic variance or common environmental effects on MZ twins that are greater than for DZ twins. Data sets with these features exemplify the need to continue studying the properties of different methods for analysis of twin data.

THE EVOLUTION OF TWIN ANALYSIS

Sir Francis Galton,[1] is credited as being the father of twin studies by proposing in an 1876 publication that twins be used to study nature versus nurture. However, over a century later there continues to be debate about what methods should be used in the analysis of twin data.

Kang et al.[2] reviewed twelve twin estimates of heritability (h^2 = the fraction of population variance due to genetic variance). Newman et al.[3] were credited with the first use of intraclass correlation (r) values of monozygotic (MZ) and dizygotic (DZ) twins to estimate h^2 with the formula $(rMZ-rDZ)/(1-rDZ)$. To be unbiased, this estimate requires that the within DZ environmental variance equals half of population environmental variance. Hancock[4] suggested using the intraclass correlation of MZ twins (rMZ) to estimate h^2, but it is not commonly used because it contains both genetic and common environmental sources of variance. However, rMZ does provide a useful upper-limit estimate of h^2. Falconer[5] stated that the formula (rMZ-rDZ) could be taken as an estimate of half of heritability if genetic effects were additive (A) and the common environmental variances (C) of MZ and DZ twins were equal. Subsequently, 2(rMZ-rDZ) has become a commonly used estimate of h^2.

THE TWIN ANALYSIS OF VARIANCE MODEL

The mean squares and their expected values in Table 9.1 are a simplified twin ANOVA model.[6] This model provides the expected values of the mean squares used in estimating and testing the significance of parameters in Table 9.2. For simplicity, only additive, two-factor (D) and three factor genetic interactions (I) are modeled, genetic-environmental interactions are not included and all population parameters (A, C, D, E and I) are assumed to be equal in MZ and DZ twins.

Partitioning Environmental Components of Twin Variance

The expected value of the within-pair MZ mean square (WMZMS) and the within-pair DZ mean square (WDZMS) both contain a component of environmental variance unique to individuals ($\sigma^2 E$). Because $\sigma^2 E$ effects are unique to individuals this component of variance is also contained in the among-pair mean squares of both zygosities. Cotwins share, to varying degrees, a common environment from conception, if the common environments influence a variable being studied, there will be a component of environmental variance common to cotwins

Table 9.1 – Analysis of Variance Model for Twin Studies

Source of Variation	Mean Squares	Expected values of mean squares
MZ Twins		
Among pairs	AMZMS	$2\sigma^2 A + 2\sigma^2 D + 2\sigma^2 I + \sigma^2 E + \sigma^2 C$
Within pairs	WMZMS	$\sigma^2 E$
DZ twins		
Among pairs	ADZMS	$3/2\, \sigma^2 A + 5/4\, \sigma^2 D + 9/8\, \sigma^2 I + \sigma^2 E + \sigma^2 C$
Within pairs	WDZMS	$1/2\, \sigma^2 A + 3/4\, \sigma^2 D + 7/8\, \sigma^2 I + \sigma^2 E$

Note: MZ, DZ = monozygotic and dizygotic twins, respectively; AMZMS, WMZMS, ADZMS, WDZMS = mean squares among and within MZ pairs and among and within DZ pairs, respectively; $\sigma^2 A$ = variance component due to additive genetic effects; $\sigma^2 D$ = variance component due to the interaction of two genetic factors (dominance or two-factor epistasis); $\sigma^2 I$ = variance component due to the interaction of three genetic factors; $\sigma^2 E$ = variance component due to environmental effects unique to individuals; $\sigma^2 C$ = environmental effects common to cotwins.

Table 9.2 – Twin Analysis of Variance (ANOVA): Parameter Estimates and Significance Tests

Parameter	Estimate	Significance Test	Comments
1. Normal Distributions	Deviation from normality	Kolmogorov-Smirnov	Transformations may remove significant deviations.
2. Δ between μ of twins 1 and 2	$|\,\overline{x}_1 - \overline{x}_2\,|$	Paired Comparison t	Used in cotwin Control Studies
3. Δ between σ^2 of twins 1 and 2	$|\,MS1's - MS2's\,|$	$F = \dfrac{\text{larger MS}}{\text{smaller MS}}$	Primary use in likelihood models
4. Δ between μ of MZ and DZ twins	$|\,MZ\,\overline{x} - DZ\,\overline{x}\,|$	t'	If significant (P<0.05) may seriously bias variance estimates
5. Δ between total MZ and DZ σ^2	$|\,\Sigma MSMZ - \Sigma MSDZ\,|$	$F = \dfrac{\text{larger }\Sigma MS}{\text{smaller }\Sigma MS}$	If P<0.2 may seriously bias most heritability estimates
6. Zygosity intraclass correlations (ρ)	$\dfrac{AMS - WMS}{AMS + WMS}$	$F = (AMS/WMS)$	Intraclass correlations are standardized covariances
a. Combined	$r(MZ + DZ)$	$F = (AMS/WMS)$	Power similar to likelihood test of A significance in AE model or C in CE model
b. Each Zygosity	$r(MZ)$ or $r(DZ)$	$F = (AMS/WMS)$	rMZ is a useful upper limit estimate of (A+C)/Total Variance

7. Twin genetic variance (G_T) A complex fraction of genetic variance including: $1/2\ \sigma^2 A + 3/4\ \sigma^2 D + 7/8\ \sigma^2 I$

Parameter	Estimate	Significance Test	Comments
a. Within pairs (G_{WT})	WDZMS-WMZMS	$F = \dfrac{WDZMS}{WMZMS}$	Biased upward if $\sigma^2 DZ > \sigma^2 MZ$
b. Among pairs (G_{AT})	AMZMS-ADZMS	$F = \dfrac{AMZMS}{ADZMS}$	Biased upward if $\sigma^2 DZ < \sigma^2 MZ$
c. Among component (G_{AC})	(AMZ+WDZMS's)–(WMZ+ADZMS's)	$F' =$ one-tailed	Unbiased if unequal zygosity total variance due to environment ($\sigma^2 E$)

8. Intraclass Correlation Estimates of Variance Components

Parameter	Estimate	Significance Test	Comments
a. Heritability (h^2)	2(rMZ–rDZ)	ZrMZ–ZrDZ	Assumes D=I=0.0
b. Common Environmental Variance (C)	2(rDZ)–rMZ	Z2(rDZ)–ZrMZ	
c. Gene Interactions (D or I)	rMZ–2(rDZ)	ZrMZ–Z2(rDZ)	Biased upward if CMZ>CDZ

Note: See also abbreviations with Table 9.1. Δ = difference; σ^2 = variance parameter; ρ = correlation parameter; t = t-test; F = F-test; t' and F' = t and F tests done using a combination of mean squares and approximate degrees of freedom; $\overline{x}$ = mean estimate; μ = mean parameter; Σ = sum; Z = Z transformation of intraclass correlations.

(σ^2C). The σ^2C component of environmental variance influences variation among pairs of twins and is therefore only contained in the expected values of the among-pair mean squares.

Partitioning Genetic Components of Twin Variance

MZ cotwins share the same genotype so the expected value of the MZ within-pairs mean square contains no genetic variance. The among-pairs mean squares are calculated from the sum of pairs of individuals, resulting in the MZ among-pairs mean square expected value containing twice the population genetic variance, irrespective of whether the genetic effects are additive (A) dominant (D) or due to interactions of three genetic factors (I). As DZ twins inherit, on average, one-half of the same genes from their parents, the probability that they will inherit one gene in common is 50%, any two genes in common is 25% and any three genes in common is 12.5%. These probabilities explain why the within-DZ mean squares (WDZMS) are expected to contain 1/2 of the population additive genetic effects (A) which are due to the cumulative effects of single genes. When gene interactions are present, a larger share of population genetic variance is expected to occur between DZ cotwins, specifically, the WDZMS contains an expected 3/4 of the population genetic effects requiring the interaction of two genes (D) and 7/8 of the genetic effects requiring the interaction of three genes (I). The DZ among-pair mean squares are calculated using the sums of two individuals, resulting in the expected values for the DZ among-pairs mean squares (ADZMS) for genetic components of variance of 3/2 σ^2A, 5/4 σ^2D and 9/8σ^2I.

TESTABLE ASSUMPTIONS OF THE TWIN MODEL

Deviations From Statistically Normal Distributions

The estimation of departures of the zygosity distributions from normality (Parameter 1, Table 9.2) is critical for most subsequent tests. A commonly used normality test is the Kolmogorov-Smirnov goodness-of-fit test for each zygosity. If the p value is less than 0.05, an appropriate transfomation from the Box-Cox series is suggested.[7] These transformations primarily remove skewness. Estimates of genetic effects using mean differences between cotwins have been shown to be more robust to departures from normality, but these estimates have not been widely used.[8, 9]

Differences Between the Means of Cotwins

Tests of the significance of the difference (Δ) between the means of twins 1 and 2 in each pair are used primarily in cotwin control studies where MZ twins are used to search for environmental influences and DZ twins may be used to search for genetic associations (Parameter 2, Table 9.2). Because of the similarities of cotwins, and the use of paired-comparison tests of significance, cotwin control studies are often more powerful than comparisons of unrelated individuals (see Chapter 5).[10] Cotwin control studies are also ideal for sequential analysis which further increases their power.[11] A tertiary method of increasing the power of cotwin control studies is to select subjects from cotwins who have similar phenotypes.

Differences between the Variances of Cotwins

The mean squares of the individual twins within zygosities are not modeled in Table 9.1, but when calculated they may be used to estimate and test for significance of differences between the variances of cotwins, using a two-tailed F-ratio (Parameter 3, Table 9.2). This comparison is not usually done in ANOVA, but is done in likelihood analyses as the variances of cotwins are estimated in the construction of the zygosity covariance matrices.

Differences between the Means of MZ and DZ Twins

The significance of the difference between the means of the two zygosities or any two types of twins may be tested using a t' test (Parameter 4, Table 9.2).[12] The t' test differs from the t test in that the denominator is constructed from two mean squares (AMZMS and ADZMS) rather that a single mean square. This is required as the expected values of these two mean squares are different (Table 9.1). At times, a t-test is used to test this difference, but this is not appropriate, and may give false positive results as it ignores the partitioning of variance estimated by the among and between twins mean squares. If there is a significant difference between the means of MZ and DZ twins, it is likely, but not a certainty, that subsequent variance estimates will be biased

Differences between the Total Variances of MZ and DZ Twins

Differences between the total variances of MZ and DZ twins will cause biases in most parameter estimates obtained by comparing the two zygosities (Parameter 5, Table 9.2). It has been recommended that if this difference has a $p < 0.2$ (two-tailed F' test), subsequent parameter estimates should be considered biased.[13] One possible exception is the among-component estimate of twin genetic variance (see G_{AC} below).

ESTIMATING GENETIC AND ENVIRONMENTAL COMPONENTS OF VARIANCE

Twin Correlations

Twin population correlations (ρ), for either zygosity or the combined zygosities may be estimated and tested for significance using intraclass correlation (r) values and their Z-transformations in ANOVA (Parameter 6, Table 9.2)[14] or using a covariance matrix in likelihood analyses. The Z-transformation, where $Z = 1/2$ ln $[(1+r)/(1-r)]$, is employed to normalize the distribution of correlation coefficients. Twin r values are estimates of standardized population covariance and may be tested for significance by the F-ratio (AMS/WMS). The intraclass correlation combining both zygosities [r(MZ+DZ)] provides a sensitive test for the presence of twin covariance, however it is seldom calculated. It is included in Table 9.2 because its power is somewhat greater than the power of the test for C in the CE vs E likelihood models. As shown in Table 9.3 r(MZ+DZ) was significant 64.1% of the time when C = 20% of the total variance and 50 pairs each of MZ and DZ twins were simulated. This compared to 52.9% of the time when the same simulation was used to compare the CE and E likelihood models. Neale and Cardon[15] recommended that if the fit of the ACE likelihood model is not significantly better than the fit of the AE model, it is appropriate to accept the null hypothesis that C=0.0 and use the AE model to estimate and test the significance of A. A problem with this recommendation is that both the comparison of the ACE and CE models as well as the comparison of rMZ and 2rDZ are both so lacking in power that relatively large components of C may go undetected in most twin studies. For example, in Table 9.4, when C was simulated to represent 40% of total variance, the ANOVA test found it significant only 28.0% of the time and the maximum likelihood test was only slightly better at 31.4%. A review of the literature has provided several examples in which estimates of heritability were apparently biased upward by the inclusion of C.[14] In fact, as C increases, the AE model is biased upward further so that it is possible to obtain estimates of H^2 markedly greater than rMZ or 1.0. Difficulty in partitioning twin covariance into A and C arises because estimate of these two components are negatively correlated.[16]

There is some disagreement as to what standard of significance should be applied to estimates of twin covariance before attempting to partition the covariance into genetic and common environmental components. Most would agree that if neither

Table 9.3 – Comparison of the power to detect twin covariance by the percentage of significant results for the ANOVA r(MZ+DZ) and the maximum likelihood comparison of the CE vs E models using 1,000 simulations of studies with 50 MZ and 50 DZ twin pairs.

	% Significant Results (p<0.05)	
C = % Variance	**r(MZ+DZ)**	**CE vs E**
10	23.8	17.0
20	64.1	52.9
30	89.8	84.6
40	99.4	98.7
50	100.0	100.0

Note – The simulations were done using values drawn randomly from normally distributed variables

Table 9.4 – Comparison of the power to detect C by the percentage of significant results for the ANOVA Z2(rDZ) – ZrMZ and the maximum likelihood comparison of the ACE vs. AE models using 1,000 simulations of studies with 50 MZ and 50 MZ twin pairs.

	% Significant Results ($p < 0.05$)	
C = % Variance	**Z2(rDZ) – ZrMZ**	**ACE vs AE**
20	7.9	9.6
40	28.0	31.4
60	62.1	80.2
80	91.0	99.7

Note – The simulations were done using values drawn randomly from normally distributed variables

zygosity has an ρ value greater than 0.0, there is little to be gained by further attempts to partition the covariance. Hopper[17] recommended that both zygosities have ρ values greater than 0.0 before attempting further variance partitioning. The argument against this restriction is that gene interactions are expected to cause small population DZ covariance, even in the presence of a major component of genetic variance. For example, it is estimated that if the population true correlation (ρ) of DZ twins is 0.125, to detect it 80% of the time ($P < 0.05$) requires over 400 pairs of DZ twins.[14] A ρ for DZ twins of 0.125 or less is expected due to gene interactions (D or I in Table 9.1) if they account for half of the population variance.

When a decision is made to estimate genetic and environmental components of variance, several assumptions remain that are usually not testable, but are required for unbiased estimates, including:

1. Mating is random in the population.
2. MZ and DZ twins are random samples of population twins.
3. There are no genetic-environmental interactions.
4. Environmental covariance (C) of MZ twins equals that of DZ twins.

Estimates of Population Genetic Variance from Twin Data by Comparing Mean Squares

Because of the differences in expected values of the MZ and DZ twin mean squares, and if the above assumptions hold, a complex fraction of population genetic variance can be estimated. There are three combinations of mean squares that can be used to estimate this fraction of population genetic variance (G_T, Parameter 7, Table 9.2). These three combinations are explained under headings a. (G_{WT}), b. (G_{AT}), and c. (G_{AC}) under parameter 7 (Table 9.2) and using the model in Table 9.1 all are expected to contain the same complex fraction of population genetic variance: [G_T estimates $(1/2 \; \sigma^2 A + 3/4 \; \sigma^2 D + 7/8 \; \sigma^2 I)$]. These estimates are expected to be equal within sampling variation, if the assumptions of the twin model hold; however, they can vary markedly if these assumptions do not hold.

G_{WT} and G_{AT} are tested for significance using one-tailed F-ratios while G_{AC} is calculated using two mean squares in both the numerator and the denominator and requires an F' test rather than a simple F test.[13] Environmental variance unique to individuals (E) may be estimated by subtracting G_T from an estimate of total population variance (D+I=0.0). These estimates of twin genetic variance may be transformed into h^2 values by assuming the D+I=0.0, doubling them and dividing them by the following estimate of total population variance [σ^2_T is estimated by (AMZMS+WDZMS+ADZMS+WDZMS)/4].

Failure of one or more of the assumptions of the twin model can cause marked differences among the three estimates of genetic variance. Perhaps the most commonly observed failure of a testable assumption is that of equal zygosity total variance. If DZ total variance is greater than MZ total variance, due to greater DZ than MZ $\sigma^2 E$ the within-pairs estimate will be biased upward and the among-pairs estimate biased downward by the same expected amount. For example, in a population where $\sigma^2 A = 40$, $\sigma^2 C = 20$ and $\sigma^2 E = 40$, but for some reason unique to DZ twins, $\sigma^2 E_{dz} = 60$, the true population h^2 would be 0.4. In this example heritability values calculated from G_{WT} would be biased upward at 0.8, G_{AT} biased downward at 0.0 and G_{AC} unbiased at 0.4. [13] In this example, h^2 calculated from the formula 2(rMZ-rDZ) would also be upwardly biased at 0.54. Even though the G_{AC} estimate would be theoretically useful in this situation, the F' significance test for G_{AC} has relatively little power to detect population genetic variance and this power decreases as the difference in total variances increases.

Intraclass Correlation Estimates of Variance Components

Intraclass correlation (r) values are used to estimate population correlations (ρ) and are commonly reported in both ANOVA and likelihood analyses, providing a common ground for comparing the two methods (Parameter 8, Table 9.2). In ANOVA analyses the intraclass correlations (r values) are calculated from the formula (AMS-WMS)/(AMS+WMS). In likelihood analyses the r values are calculated using an estimate of twin covariance calculated from a covariance matrix in which the variances of each zygosity are divided into the variance of twins 1 and 2. Individual twins are assigned as twin 1 or twin 2 either randomly or based upon some prior information. The estimate 8.a, in Table 9.2, calculated from the 2(rMZ-rDZ), has the same expected value of h^2 values calculated from the three genetic variance estimates above if the assumptions of the twin model hold. This estimate has slightly less power than the likelihood h^2 values obtained and tested by

Table 9.5 – Comparison of the power to detect A by the percentage of significant results for the ANOVA ZrMZ – ZrDZ and the maximum likelihood comparison of the ACE vs. CE models using 1,000 simulations of studies with 50 MZ and 50 DZ twin pairs.

A = % Variance	% Significant Results (p<0.05)	
	ZrMZ – ZrDZ	ACE vs CE
20	8.6	8.5
40	20.5	21.5
60	46.2	54.4
80	91.4	97.3

Note – The simulations were done using values drawn randomly from normally distributed variables

comparing the ACE and CE likelihood models as shown in Table 9.5. Differences between the estimates obtained from the two methods and their power may occur due to variation between the mean squares of twins 1 and 2, within zygosities.

If common environmental effects (C) are present and equal for MZ and DZ twins, they are expected to increase rMZ and rDZ equivalent amounts. Therefore if all genetic variance is additive, the difference (2rDZ – rMZ) estimates C. Simulation studies reveal that this ANOVA estimate of C is somewhat less powerful than the estimate obtained by comparing the ACE and AE likelihood models (Table 9.4).

When rMZ is greater than twice rDZ, the difference ZrMZ-Z2(rDZ) may be used to test for the significance of genetic variance attributable to gene interactions. Both the ANOVA model and the likelihood comparison of the ADE and AE likelihood models have relatively little power to detect D, although the ANOVA model was somewhat more powerful at the higher levels of D. For example, if D made up 80% of the total variance, the ANOVA test would be significant 54.9% of the time and the likelihood ratio test significant 34.1% of the time (Table 9.6). If the assumption that the common environmental effects on MZ and DZ twins are equal does not hold and CMZ>CDZ, this test for gene interactions may produce false positive results. An empirical test to screen for the possibility of either gene interactions or CMZ>CDZ is to compare the h^2 values obtained to rMZ and if h^2 is larger than rMZ this indicates the possibility of either CMZ>CDZ or gene interactions.[18] If there is evidence for gene interactions, twin covariance may be partitioned into A and D or D and I.[6] This partitioning is based upon the expectations:

1. D is expected to increase the covariance of MZ twins four times that of DZ twins and;
2. I is expected to increase the covariance of MZ twins eight times that of DZ twins.

Table 9.6 – Comparison of the power to detect D by the percentage of significant results for the ANOVA ZrMZ – Z2(rDZ) and the maximum likelihood comparison of the ADE vs. AE models using 1,000 simulations of studies with 50 MZ and 50 DZ twin pairs.

	% Significant Results ($p < 0.05$)	
D = % Variance	ZrMZ – Z2(rDZ)	ADE vs AE
20	5.3	6.6
40	9.1	9.1
60	20.6	17.7
80	54.9	34.1

Note – The simulations were done using values drawn randomly from normally distributed variables

EXAMPLES OF FAILURES OF TWIN MODEL ASSUMPTIONS

Unequal Zygosity Total Variances

Body Mass Index

In 1969 the National Heart, Lung and Blood Institute (NHLBI) began a twin study of heart disease risk factors.[19] A cohort of 514 white, male, veteran twins (250 MZ and 264 DZ pairs) ascertained from the NAS-NRC Twin Panel. Body mass index [BMI= weight in (kilograms)/height in (meters)^2], was measured, during three examinations (twins aged 42–56, 53–64 and 59–70). Between examinations one and two, the MZ twins developed a significantly smaller BMI total variance ($p=0.004$) than the DZ twins, and this difference persisted at examination three ($p= 0.003$).[20] In retrospect, it was found that MZ twins with extremes of BMI had failed to return to examinations 2 and 3 in greater numbers than expected by chance alone. When the induction records of this cohort were examined (ages 18–28) it was found that the MZ twins who returned to examinations 2 and 3 already had a smaller BMI total variance than the DZ twins or the entire population of MZ twins at induction. In this example, unequal zygosity total variance appears to have arisen because of self-selection.

High Density Lipoprotein Cholesterol

When total plasma cholesterol data were analyzed from examination one of the NHLBI Twin Study the total variance of DZ twins was 1.4 times that of MZ twins ($p<0.001$).[21] Depending upon the ANOVA estimate, genetic variance ranged from 0.4 times total variance to negative values. The greater DZ than MZ total variance persisted for 16 years at the next two examinations. At three examinations (1970,1980, and 1986) the majority of the difference between zygosity total variances was found to be due to the high-density lipoprotein (HDL) fraction of cholesterol.[22] At the third examination, the HDL cholesterol was divided into the HDL_2 (density = 1.063 to 1.125 g/ml) and HDL_3 (density = 1.125 to 1.210) fractions. The HDL_2 fraction was found to be the primary cause of the greater DZ total variance. Reed et al.[23] found that this difference in HDL total variance was associated with dermatoglyphic patterns and a dermatoglyphic index of placental type in these aging male twins.

Dermatoglyphic Patterns

In a study of 71 dermatoglyphic variables, 21 had significantly different ($p < 0.05$) zygosity total variances.[24] Because the dermatoglyphic patterns are established by the 20th week of gestation, these differences in total variance led to the study of the relationship between placental type and dermatoglyphic patterns in MZ twins.[25] Of 84 variables studied 5 had significant ($p < 0.05$) differences between the means of monochorionic-MZ (MC-MZ) and dichorionic-MZ (DC-MZ) twins, not necessarily different than expected by chance. However, all five of these variables were related to placement of the most distal axial triradius. Comparison of the chorion-type within-pair mean squares is a more sensitive test for differences between the environmental covariances of twin types than total variance differences. Of the same 84 dermatoglyphic variables, 19 had significant ($p < 0.05$) differences between the within-pair mean squares with the DC-MZ within-pair mean squares greater in 11 and the MC-MZ mean square larger in 8 of these comparisons. These results led to the development of the dermatoglyphic index of placental type within MZ twins.[26]

Relationships Between Cognitive and Personality Development and Placental Type

The dermatoglyphic index of placental type used to study HDL-C was also used to study type-A personality in the NHLBI Twin Study of World War II veteran twins.[26] Seven measures of type A personality were measured in 123 MZ twin pairs. For all seven measures there was a correlation between the placental index within-pair differences and the type A scales within-pair differences. For three of the seven measures, the correlations were significant ($p < 0.05$). In contrast, measures of hostility and cognitive function were not found to be related to placental type in this study.

Karras et al.[27] studied a cohort of MZ twins who had their placental types ascertained at birth. They found within-pair differences significantly ($p < 0.05$) smaller for the monochorionic than the dichorionic MZ pairs for three of four Personality Inventory for Children (PIC) factor scales, 8 of 12 clinical scales and 2 of 4 validity/screening scales. In contrast, no consistent relationship between placental type and cognitive function was found using the McCarthy Scales of Cognitive Ability.

The placental membranes were examined from a cohort of twins born in Toronto between 1936 and 1939. A subset of these twins were studied (17 MC-MZ and 15 DC-DZ) in the early 1980s. Two subtests of the Wechsler Adult Intelligence Scale (WAIS), Vocabulary and Block Designs, were administered. Even with this small sample, there was a significant association of the Block Design results and the within-pair mean squares with placental type (DC>MC).[28]

A cohort of twins of known placental type ascertained in the National Institute of Neurological and Communicative Disorders and Stroke (NINCDS) Collaborative Perinatal Project (NCPP), were administered the Wechsler Intelligence Test for Children (WISC) at age 7. Melnick et al.[29] reported that within the white twin pairs (23 MC-MZ and 9 DC-MZ) the within-pair mean square was significantly larger (P<0.01) for the DC-MZ than the MC-MZ. Black twins (30 MC-MZ pairs and 24 DC-MZ pairs) followed the same trend, but the association was not significant.

All of these studies indicate that there are failures of the assumptions of the twin model relating to equal total variances or equal common environmental variances for the two zygosities.

Failure of the Additive Genetic Model for Central Nervous System Function

Lykken et al.,[30] in adult male twins, reported that for total EEG spectrum power, the rMZ value of 0.81 was judged too much larger than the rDZ value of 0.12 to be readily explained by additive genetic influences alone. They repeated their study [31] and found similar results. Lykken and coworkers did not statistically test their results, but retrospectively Christian et al.[6] found, using their data, that rMZ was significantly larger than 2(rDZ), in spite of the relatively poor power of this test. In a similar study of both male and female adult twins,it was found that the correlations for total spectrum power (0.1–20 Hz) and beta-band power (13.1–20.0 Hz) were very similar to those reported by Lykken et al. (total spectrum rMZ=0.70 and rDZ 0.12) and rMZ was significantly (p<0.01) greater than 2(rDZ). Beijsterveldt et al. [32] studied EEG power in adolescent twins and found evidence for dominance of beta power at two electrodes. In addition, in their comparisons of male and female MZ and DZ twins, they found in 22 of 28 comparisons that rMZ was greater than 2rDZ. Van Baal et al.[33] studied the EEG power spectrum in 5 year old MZ and DZ twins. They did not include dominance in their models, but similar to the results of Beijsterveldt et al.[32] they found that in 35 of 56 comparisons for relative beta wave power the rMZ values were greater than 2rDZ.

In a study of auditory event-related potentials (ERP), O'Connor et al.[34] found that the amplitude of the P3 component of the ERP had an rMZ, averaged over 16 electrodes, of 0.510 compared to an rDZ which averaged -0.036 (rMZ>0.0 and 2rDZ,

P<0.001). Boomsma and Gabrielli [35] (1985) reviewed the literature on twin studies of visual evoked potentials and concluded that rMZ values were around 0.50 while rDZ values were near 0.0, a finding compatible with gene interactions or CMZ>CDZ.

Gale et al. [36] et al. studied saccadic eye movement latency and accuracy in adult twins before and after alcohol administration. All 12 rMZ values calculated were greater than 2rDZ, suggesting that the additive genetic model was not adequate to explain these findings. The most likely causes of rMZ values apparently too large, compared to rDZ values, to be attributed to additive genetic influences, would appear to be either gene interactions, as postulated by Lykken et al.,[30, 31] or due to greater environmental covariance for MZ than DZ twins (CMZ>CDZ).[6] This finding appears commonly in electrical measurements of brain activity and, in the case of eye movements, physical activity closely related to central nervous system function. In conclusion, the various assumptions of twin analysis can in most cases be formally tested, although the power of twin studies to detect common environmental effects and dominance is low in most studies. Examples of unequal variance in twins and selection bias show the need to continue to study the performance of different estimators of genetic variance and heritability under a variety of conditions.

REFERENCES

1   Galton F. The history of twins as a criterion of the relative powers of nature and nurture. *J Anthrop Inst Great Brit and Ireland 1876*; **5**: 391–406

2   Kang KW, Christian JC, Norton JA Jr. Heritability estimates from twin studies. *Acta Genet Med Gemellol* (Roma) 1978; **27**: 39–44

3   Newman HH, Freeman FN, Holzinger JJ. *Twins: A study of heredity and environment*. Chicago: University of Chicago Press. 1937

4   Hancock J. Studies in monozygotic twins. *New Zealand Science and Technology* 1952; **34**: 131–135

5   Falconer DS. *Introduction to Quantitative Genetics*. New York: The Ronald Press, 1960: p 185

6   Christian JC, Morzorati S, Norton JA Jr, Williams CJ, O'Connor S, Li T-K. Genetic analysis of the resting electroencephalographic power spectrum in human twins. *Psychophysiol* 1996; **33**: 585–591

7   Williams CJ, Christian JC, Norton JA Jr. TWINAN90: A FORTRAN program for conducting ANOVA-based and likelihood-based analyses of twin data. *Comput Methods Programs Biomed* 1992; **38**: 167–176

8   Bailey-Wilson JE. The effects of failures of assumptions on several tests used for genetic analysis. Indiana University Ph.D. Thesis 1981

9 Williams CJ, Zhou L. A comparison of tests for detecting genetic variance from human twin data based on absolute intratwin differences. *Communications in Statistics: Simulation and Computation*; 1998; **27**: 51–65

10 Christian JC, Kang K-W. Efficiency of human monozygotic twins in metabolic studies of blood lipids. *Metabolism* 1972; **21**: 691–699

11 Christian JC, Yu PL, Grim CE, Norton JA Jr. The use of sequential analysis to improve the efficiency of cotwin control studies. *Acta Genet Med Gemellol* (Roma) 1980; **29**: 163–165

12 Christian JC, Norton JA Jr. A proposed test of the difference between the means of monozygotic and dizygotic twins. *Acta Genet Med Gemellol* (Roma) 1977; **26**: 49–53

13 Christian JC, Kang K-W, Norton JA Jr. Choice of an estimate of genetic variance from twin data. *Am J Hum Genet* 1974; **26**: 154–161

14 Christian JC, Norton JA Jr, Sorbel J, Williams CJ. Comparison of analysis of variance and maximum likelihood based path analysis of twin data: Partitioning genetic and environmental sources of covariance. *Genet Epidemiol* 1995; **12**: 27–35

15 Neale MC, Cardon LR. *Methodology for Genetic Studies on Twins and Families*. Boston: Kluwer Academic Publishers. 1992

16 Williams CJ. On the covariance between parameter estimates in models of twin data. *Biometrics* 1993; **49**: 557–568

17 Hopper JL. The epidemiology of genetic epidemiology. *Acta Genet Med Gemellol* (Roma) 1992; **41**: 261–273

18 Slemenda CW, Christian JC, Williams CJ, Norton JA Jr, Johnston CC Jr. Genetic determinants of bone mass in adult women: A reevaluation of the twin model and the potential importance of gene interaction on heritability estimates. *J Bone Miner Res* 1991; **6**: 561–567

19 Feinleib M, Havlik RJ, Kwiterovich PO, Tillotson J, Garrison RJ. The National Heart Institute Twin Study. *Acta Genet Med Gemellol* (Roma) 1970; **19**: 243–247

20 Selby JV, Reed T, Newman B, Fabsitz RR, Carmelli D. Effects of selective return on estimates of heritability for body mass index in the National Heart, Lung, and Blood Institute Twin Study. *Genet Epidemiol* 1991; **8**: 371–380

21 Christian JC, Feinleib M, Hulley SB, et al. Genetics of plasma cholesterol and triglycerides: A study of adult male twins. *Acta Genet Med Gemellol* (Roma) 1976; **25**: 145–149

22 Christian JC, Carmelli D, Castelli WP, Fabsitz R, Grim CE, Meaney FJ, Norton JA Jr, Reed T, Williams CJ, Wood PD. High density lipoprotein cholesterol: A 16-year longitudinal study in aging male twins. *Arteriosclerosis* 1990; **10**(6): 1020–1025

23 Reed T, Christian JC Wood PD, Schaefer EJ. Influence of placentation on high density lipoproteins in adult males: The NHLBI twin study. *Acta Genet Med Gemellol* 1991; **40**: 353–359

24 Reed T, Sprague FR, Kang KW, Nance WE, Christian JC. Genetic analysis of dermatoglyphic patterns in twins. *Hum Hered* 1975; **25**: 263–275

25 Reed T, Uchida IA, Norton JA Jr, Christian JC. Comparisons of dermatoglyphic patterns in monochorionic and dichorionic monozygotic twins. *Am J Hum Genet* 1978; **30**: 383–391

26 Reed T, Carmelli D, Rosenman RH. Effects of placentation on selected type A behaviors in adult males in the National Heart, Lung, and Blood Institute (NHLBI) twin study. *Behav Genet* 1991; **21**(1): 9–19

27 Sokol DK, Moore CA, Rose RJ, Williams CJ, Reed T, Christian JC. Intrapair differences in personality and cognitive ability among young monozygotic twins distinguished by chorion type. *Behav Genet* 1995; **25**(5): 457–466

28 Rose RJ, Uchida IA, Christian JC. *Placentation effects on cognitive resemblance of adult monozygotes.* Twin Research 3: Intelligence, Personality and Development. New York: Alan R. Liss. 1981: pp 35–41

29 Melnick M, Myrianthopoulos NC, Christian JC. The effects of chorion type on variation n IQ in the NCPP twin population. *Am J Hum Genet* 1978; **30**: 425–433

30 Lykken DT, Tellegren A, Thorkelson KA. Genetic determination of EEG frequency spectra. *Biol Psychol* 1974; **1**: 245–259

31 Lykken DT, Tellegren A, Iaconno WG. EEG spectra in twins: Evidence for a neglected mechanism of genetic determination. *Phsysiol Psychol* 1982; **10**: 60–65

32 Van Beijsterveldt CE, Molenaar PC, De Geus EJ, Boomsma DI. Heritability of human brain functioning as assessed by electroencephalography. *Am J Hum Genet* 1996; **58**: 562–573

33 Van Baal GC, De Geus EJ, Boomsma DI. Genetic architecture of EEG power spectra in early life. *Electroenceph Clin Neurophysiol* 1996; **98**: 502–514

34 O'Connor S, Morzorati S, Christian JC, Li T-K. Heritable features of the auditory oddball event-related potential: Peaks, latencies, morphology and topography. *Electroenceph Clin Neurophysiol* 1994; **92**: 115–125

35 Boomsma DI, Gabrielli WF Jr. Behavioral genetic approaches to psychophysiological data. *Psychophysiol* 1985; **22**: 249–260

36 Gale BW, Abel LA, Christian JC, Sorbel J, Yee RD. Saccadic characteristics of monozygotic and dizygotic twins before and after alcohol administration. *Invest Ophthalmol Vis Sci* 1996; **37**: 339–344

10

PATH ANALYSIS OF AGE-RELATED DISEASE TRAITS

Harold Snieder

ABSTRACT

As many disease traits show a strong association with age, accounting for age in the quantitative genetic modelling of these traits is important. Incorporation of age into the path model prevents potential confounding of variance component estimates. Failure to allow for the association between age and the disease trait can spuriously introduce a shared environmental effect. Further, several hypotheses concerning the age-dependency of genetic influences on disease traits can be tested using path modelling to both longitudinal and cross-sectional twin data. First, the magnitude of the genetic influence on the disease trait can differ with age. Second, different genes may affect the phenotype at different ages, i.e.; the expression of genes may depend on age (or age-related events like the menopause). A range of models is applied to cross-sectional data on total body bone mineral density (BMD) from a large sample of female adult twins to illustrate the inclusion of age in the path analysis.

It is increasingly recognised that common chronic diseases like cardiovascular disease and osteoporosis have a complex etiology, i.e. multiple genetic and environmental influences determine the eventual disease outcome via their influence on intermediate disease traits. Risk for these chronic diseases is not static but changes over time, as illustrated by the strong association between age and (continuously distributed) intermediate disease traits like blood pressure and BMD.

The main aim of this chapter is to show the importance of accounting for age in the quantitative genetic analysis of age-related disease traits. It will be shown that failure to allow for the association between age and the disease trait in the analysis can confound the variance components estimates introducing a spurious shared environmental effect. Next, it is discussed how path analysis can be used to address hypotheses concerning the age dependency of genetic (or environmental) influences on disease traits. This age dependency can take two different forms. Firstly, the magnitude of the genetic influence on the disease trait can differ with age. Secondly, different genes may affect the phenotype at different ages, i.e.; the expression of genes may depend on age (or age-related events like the menopause). Although path analysis of longitudinal twin data would be ideal to solve those questions, in some cases modelling of cross-sectional twin data can also provide the answers.

A number of examples using cross-sectional data on total body BMD in a large sample of female twins (age range: 18–75 years) will be given to illustrate pitfalls and scope of quantitative genetic model fitting of age-related disease traits.

PATH ANALYSIS OF TWIN DATA

Even in the absence of genotypic information, genetic and environmental sources of individual differences in disease traits can be quantified by applying path analysis to twin or family data. In the classic twin method, the difference between intraclass correlations for monozygotic twins and those for dizygotic twins is doubled to estimate heritability $[h^2=2(r_{MZ} - r_{DZ})]$,[1] with the remaining population variance attributed to environmental factors. Estimates of genetic and environmental effects based on comparisons of intraclass correlations, however, have low power and large standard errors and do not make use of information available in variances and covariances. Path analysis (or structural equation modelling) involves solving a series of simultaneous structural equations in order to estimate genetic and environmental parameters that best fit the observed twin (or family) variances and covariances (See Box). Path analysis is much more flexible in its applications than the classic twin methodology. Hypotheses of interest can be explicitly tested by comparing the fit of alternative models and analyses can be extended to more than two twin groups, other family members, multiple variables and multiple time points[2].

AGE AS CONFOUNDER

Many disease traits show a strong association with age. Blood pressure rises with age, for example, whereas BMD shows an age-related decline. Our own data, a large sample of adult female twins from the St. Thomas' UK Adult Twin Registry (Table 10.1) showed a correlation between age and total body BMD of -0.40 [see[3,4] for recruitment and measurement procedures].

When the age range of the twin sample is broad and there is a significant correlation between age and the phenotype, age should be incorporated in the path model. Any age difference between the monozygotic (MZ) and dizygotic (DZ) twins [as in this specific example (Table 10.1)] provides an additional reason for incorporating age in the analysis. One way to do this is to model age as a linear regression (Figure 10.2), which at the same time quantifies the percentage of variance of the phenotype that is explained by age. Table 10.2 shows model fitting results of total body BMD in 378 MZ and 829 DZ female twin pairs for a model without age and one including age. Both standardised and non-standardised estimates of variance components of the best fitting models are shown. All model fitting reported in this chapter was done in Mx[5] (see also Chapter 18).

Comparison of the model fitting results for the model without age and the one

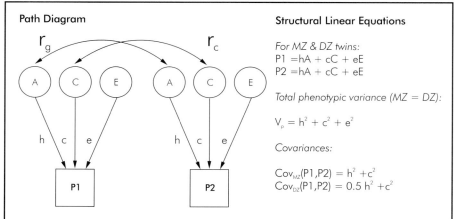

Path Diagram

Structural Linear Equations

For MZ & DZ twins:
P1 =hA + cC + eE
P2 =hA + cC + eE

Total phenotypic variance (MZ = DZ):

$V_p = h^2 + c^2 + e^2$

Covariances:

$\text{Cov}_{MZ}(P1,P2) = h^2 + c^2$
$\text{Cov}_{DZ}(P1,P2) = 0.5\, h^2 + c^2$

Figure 10.1 – On the left the path diagram for the simple twin model is shown. This path diagram can be translated into linear structural equations (right), i.e. they are equivalent representations of the same twin model.

A number of conventions are used in path analysis and the representation of the path diagram. Observed variables (for twin 1 and twin 2) are shown in squares. Latent variables (or factors) are shown in circles. A single headed arrow indicates a direct influence of one variable on another, its value represented by a path coefficient (comparable to a factor loading). Double headed arrows between two variables indicate a correlation without any assumed direct relationship. Explanation of symbols: A = additive genetic latent factor, C = shared (or common) environmental latent factor, E = unique environmental latent factor, h = (additive) genetic factor loading, c = shared (or common) environmental factor loading, e = unique environmental factor loading, P1 (P2) = phenotypic value of twin 1 (twin 2), r_g = genetic correlation (1 for MZ and 0.5 for DZ twins), r_c = shared environmental correlation (1 for both MZ and DZ twins), V_p = total phenotypic variance, h^2 = heritability, c^2 = shared environmental variance, e^2 = unique environmental variance

including age showed that about half of the shared environmental variance could actually be attributed to the effect of age. Failure to account for age can thus spuriously introduce a common environmental effect. Purely because both twins in a pair have the same age, MZ as well as DZ correlations will become inflated if the phenotype of interest shows a significant correlation with age. This age effect will show up as an effect of shared environment if not accounted for in the model fitting.[2,6]. Although our example serves the purpose of illustrating a spurious effect of shared environment if age is not included in the modelling, it may not be the best example as the relationship between total body BMD and age is known to be

Table 10.1 – Characteristics of the twin sample and the intraclass correlations for total body BMD

	MZ	DZ
N	378	829
Mean age (y)	51.4	48.0
Mean total body BMD	1.09	1.13
ICC for total body BMD	0.90	0.58

MZ = monozygotic, DZ = dizygotic, N = number of pairs, total body BMD = whole body bone mineral density, ICC = intraclass correlation

Table 10.2 – Model fitting results for total body BMD

	Without Age		With Age	
	non-stand.	stand.	non-stand.	stand.
V_A	7.75	0.62	7.68	0.61
V_C	3.45	0.27	1.72	0.14
V_E	1.36	0.11	1.36	0.11
Age	—	—	1.84	0.14
V_P	12.56	1.00	12.60	1.00

non-stand. = non-standardised, stand. = standardised, V_A = additive genetic variance, V_C = shared environmental variance, V_E = unique environmental variance, Age = variance attributed to age, V_P = total phenotypic variance.

non-linear. An error will thus be made if age is incorporated as a *linear* regression in the path model in cases where the relation between age and the phenotype is non-linear. More comprehensive examples of quantitative genetic analyses that account for the non-linear BMD-age relationship will be given later in this chapter.

AGE-DEPENDENT GENETIC EFFECTS

Not only the means but also the variances of intermediate disease traits have been found to vary with age. Variances of blood pressure ,[7] lipids and lipoproteins[8,9] have been shown to increase with age. This observation implies that the underlying genetic and/or environmental sources of the phenotypic variance must increase as well.

Another observation indicative of an age effect on disease traits and documented for systolic and diastolic blood pressure[10] and total cholesterol levels[11] is the lower parent-offspring correlation compared to sib-pair and DZ twin pair correlations (see Table 10.3). This difference in correlations is also the main reason for the peculiar finding that heritability estimates derived from family studies are generally lower than those derived from twin studies. The influence of dominance variation

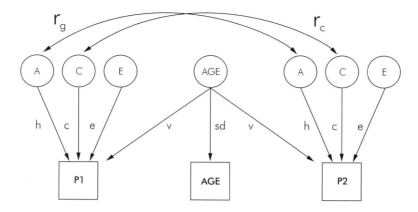

Figure 10.2 – Path model including age regression. v = factor loading on age, sd = standard deviation of age. For other symbols see Figure 10.1

may be one explanation for the difference between parent-offspring and sib- and DZ twin pair correlations. However, an effect of dominance is hardly ever found for traits like blood pressure and cholesterol. A more likely explanation, therefore, is an effect of age. One possibility is that effects of unique environmental influences accumulate over a lifetime. Alternatively, different genes may influence the disease trait in different periods of life. Both effects would explain the lower parent-offspring correlation.

The above observations can be summarised into two main questions regarding age dependency of genetic effects that can be addressed using path analysis: (i) is the *magnitude* of the genetic (or environmental) influences on the disease trait dependent on age? and (ii) do *different genes* affect the disease trait at different ages? (i.e., is there age-dependent gene expression?). Path analysis of longitudinal twin data would be the preferred method to solve those questions, but in some cases modelling of cross-sectional twin data can also provide the answers.

Longitudinal data

If data on twins have been collected on two separate occasions a triangular or Cholesky decomposition can be used to analyse the data[12]. Such an analysis can separate new genetic and environmental effects specific to the second measurement from effects that are common to the two time points. If variables have been measured on more than two occasions more sophisticated time series analysis can be applied to the data.[13] These types of analyses allow conclusions as to what extent genetic and environmental influences are the same or different across different ages (See [14], for a time series analysis of blood pressure).

Table 10.3 – Comparison between parent-offspring, sib-pair and DZ twin pair correlations for systolic blood pressure (SBP), diastolic blood pressure (DBP) and cholesterol (adapted from 10 and 11)

PAIRS	N	SBP	DBP	N	cholesterol
Pa-of	14,553	0.17	0.14	4716	0.26
Sib				2056	0.34
	11,839	0.24	0.20		
DZ				622	0.44

N = number of pairs, Pa of = Parent - offspring

Cross-sectional data

Testing for changes in magnitude of genetic influences: Stratification in age groups

Collection of longitudinal twin data is costly and time consuming and therefore not always feasible. However, path analysis of cross-sectional twin data can also offer some insight. One way to address the question whether the magnitude of the genetic influence is dependent on age is to stratify the available data into two or more age groups, which we did for our data on total body BMD in female twins.

BMD in females is (near) maximal (the PBM = Peak Bone Mass) between the age of 20 and 45 after which it starts to decline with an even sharper decline around the time of the menopause.[15] This general pattern was confirmed in our own data (data not shown). Based on this information we divided our twins into two age group with 45 years as a cut-off point. In this way the BMD of the younger group represented PBM values, whereas those from the older group represented PBM minus a certain amount of bone loss. Within each age group age was modelled as a linear

Table 10.4 – Model fitting results for total body BMD in two age groups: <45 yr ($N^{MZ}=94$; $N^{DZ}=289$) and >45 yr ($N^{MZ}=284$; $N^{DZ}=540$)

	< 45 yr		> 45 yr	
	non-stand.	stand.	non-stand.	stand.
V_A	5.83	0.71	9.55	0.74
V_C	1.43	0.17	0.00	0.00
V_E	0.94	0.12	1.48	0.11
Age	0.00	0.00	1.90	0.15
V_P	8.21	1.00	12.93	1.00

Note: non-significant variance components are fixed to zero
For abbreviations see Table 10.2.

regression. Model fitting results of the two age groups are shown in Table 10.4. As has been found for other disease traits, the total phenotypic variance is larger in the older age group. This is also reflected in the larger underlying genetic and environmental variances in this age group. As expected, age does not contribute significantly to BMD in the younger age group (i.e., BMD remains constant over this age range), but explains 15% of the variance in BMD in the older group, signifying an age-related decrease in BMD after the age of 45. A contribution of the shared environment becomes non-significant in the older age group. This indicates that the influence of the shared familial environment (for example, calcium in the diet and physical exercise) loses its effect on bone in later life (see also Chapter 13). In conclusion, this analysis shows an increase with age in the *magnitude* of genetic and environmental influences on total body BMD.

Instead of comparing variance components estimates between two or more age groups an alternative way to investigate age effects is to use curve-fitting procedures, in which genetic and environmental variance components are modelled as a continuous function of age. See[16] for an example of such an analysis applied to BMD in 10 to 26 year old females.

Testing for age-dependent gene expression: use of an extended model

In some specific cases cross-sectional data can be used to test whether the expression of genes is dependent on age in addition to testing for differences in magnitude of genetic influences. One way to do this is to use an extended model that combines data from middle-aged parents and their adolescent twin offspring with data from twins of the same age-range as those parents.[7,9] Such an extended quantitative genetic model allows testing if and to what extent different genes influence a phenotype in adolescence as compared to adulthood.

Testing for age-dependent gene expression: use of a gene × environment interaction model

For the analysis of BMD and bone loss it can be argued that rather than using a more or less arbitrary cut-off point of 45 years (a gene × age interaction model) it would make more biological sense to use information on menopausal status and treat the occurrence of menopause as a cut-off point in the analysis (a gene × menopause interaction model). Such a gene × menopause interaction model can be regarded as a gene × environment interaction model [see [2], pp. 223–229] in the sense that the immediate environment of female bone alters significantly by the sharp drop in blood concentration of oestrogen during the menopausal transition.

In this model the total sample can be subdivided into six menopause by zygosity groups. Table 10.5 shows the twin correlations for total body BMD in these six groups. In addition to estimating the magnitude of genetic influences on total body BMD in pre- and postmenopausal women it can be tested whether BMD is influenced by different genes before and after the menopause. To answer the latter question information from the menopause discordant groups is essential. The correlation between the latent genetic factors in these pairs (denoted by r_g, see Box) can be estimated and it can be tested whether it is lower than the normal value of 1 in MZ pairs and 0.5 in DZ pairs, which would indicate that (partly) different genes influence total body BMD in pre- and postmenopausal women.

Variance component estimates for pre- and postmenopausal women of the best fitting model are shown in Table 10.6. The pattern is similar to the results of the gene × age model as shown in Table 10.4: the total phenotypic variance is larger in the postmenopausal group which is again also reflected in a larger genetic and environmental variance. Interestingly, the standardised genetic variance component, i.e. the heritability, shows exactly the opposite effect. It decreases from 0.88 in premenopausal to 0.80 in postmenopausal women. This illustrates that conclusions based on the standardised variance components only can be highly misleading. To prevent invalid conclusions information from non-standardised estimates should, therefore, always be taken into account when judging age effects. As opposed to the younger age group in the gene × age model (Table 10.4), the

Tabel 10.5: Twin correlations for the six menopause by zygosity groups

	N	twin correlation
MZ pre/pre	99	0.86
DZ pre/pre	286	0.54
MZ pre/post	9	0.92
DZ pre/post	73	0.40
MZ post/post	243	0.88
DZ post/post	332	0.52

pre = premenopausal, post = postmenopausal. For other abbreviations see Table 10.1

Table 10.6: Model fitting results of the gene × menopause interaction model for total body BMD

	pre-menopausal		post-menopausal	
	non-stand.	stand.	non-stand.	stand.
V_A	7.54	0.88	09.94	0.80
V_C	0.00	0.00	0.00	0.00
V_E	0.98	0.12	1.46	0.12
Age	0.00	0.00	1.03	0.08
V_P	8.52	1.00	12.43	1.00

Note: non-significant variance components are fixed to zero
For abbreviations see Table 10.2

influence of shared environment did not reach significance in the premenopausal group. In the full model Vc was estimated to be 15%, but dropping Vc from the model did not lead to a significant deterioration of fit. Probably an indication of the limited power to detect an effect of shared environment in the twin design (see chapters 9 & 13). Age only contributes significantly in the postmenopausal group explaining 8% of the variance. Estimating the correlation between the latent genetic factors for menopause discordant pairs (i.e., r_g) showed no evidence for different genes influencing total body BMD in pre- and postmenopausal women. Given the small number of menopause discordant pairs, which is probably a consequence of the fact that timing of the menopause itself is under genetic influence,[17] and corresponding limited power to resolve alternative hypotheses, ideally these results need to be confirmed by a longitudinal twin or family study of change in BMD.

CONCLUSIONS

This chapter examined several ways in which age-related disease traits can be analysed using path analysis. For traits that show a linear relationship with age, incorporation of age as a linear regression into the path model offers a solution that prevents potential confounding of variance component estimates. Failure to allow for the association between age and the disease trait in the model can spuriously introduce a shared environmental effect if the age range of the twin sample is broad.

Furthermore, hypotheses concerning the age-dependency of genetic influences on disease traits can be tested using path modeling: (i) the magnitude of the genetic influence on the disease trait can differ with age and (ii) different genes may affect the phenotype at different ages. Path analysis applied to longitudinal twin data would be the preferred method. However, in less ideal circumstances when only cross-sectional twin data are available age dependency of genetic influences can still be tested using the appropriate path modelling. It was shown that stratification in age groups can resolve the first hypothesis, whereas a gene by menopause interaction model can provide the answers to both questions.

REFERENCES

1   Falconer DS. *Introduction to quantitative genetics (3rd ed.)*. Harlow: Longman, 1989

2   Neale MC, Cardon LR. *Methodology for genetic studies of twins and families.* Dordrecht: Kluwer, 1992

3   Spector TD, Keen RW, Arden NK, Morrison, NA, Major PJ, Nguyen TV, Kelly PJ, Baker JR, Sambrook PN, Lanchbury JS, Eisman JA. Influence of vitamin D receptor genotype on bone mineral density in postmenopausal women: a twin study in Britain. *Br Med J* 1995; **310**: 1357–1360

4 Arden NK, Baker J, Hogg C, Baan K, Spector TD. The heritability of bone mineral density, ultrasound of the calcaneus and hip axis length: A study of postmenopausal twins. *J Bone Miner Res*, 1996, **11**, 530–534

5 Neale MC. *Mx: Statistical modeling*. Box 710 MCV, Richmond, VA 23298: Department of Psychiatry. 2nd edition, 1997

6 Neale MC, Martin NG. The effects of age, sex, and genotype on self-report drunkenness following a challenge dose of alcohol. *Behav Genet* 1989; **19**: 63–78

7 Snieder H, van Doornen LJP, Boomsma DI. Genetic developmental trends in blood pressure levels, and blood pressure reactivity to stress. In: Turner JR, Cardon LR, Hewitt JK, eds. *Behavior genetic approaches in behavioral medicine*. New York: Plenum, 1995: pp 105–130

8 Reilly SL, Kottke BA, Sing CF. The effects of generation and gender on the joint distributions of lipid and apolipoprotein phenotypes in the population at large. *J Clin Epidemiol* 1990; **43**: 921–940

9 Snieder H, van Doornen LJP, Boomsma DI. The age dependency of gene expression for plasma lipids, lipoproteins, and apolipoproteins. *Am J Hum Genet* 1997; **60**: 638–650.

10 Iselius L, Morton NE, Rao, DC. Family resemblance for blood pressure. *Hum Hered* 1983; **33**: 277–286

11 Iselius L. Analysis of family resemblance for lipids and lipoproteins. *Clin Genet* 1979; **15**: 300–306

12 Loehlin JC. The Cholesky approach: A cautionary note. *Behav Genet* 1996; **26**: 65–69

13 Boomsma DI, Molenaar, PCM. The genetic analysis of repeated measures I: Simplex models. *Behav Genet* 1987; **17**: 111–123

14 Colletto GM, Cardon LR, Fulker DW. A genetic and environmental time series analysis of blood pressure in male twins. *Genet Epidemiol* 1993; **10**: 533–538

15 Compston JE. Osteoporosis. *Clin Endocrinol* 1990; **33**: 653–682

16 Hopper JL, Green RM, Nowson CA, Young D, Sherwin AJ, Kaymackci B, Larkins RG, Wark JD. Genetic, common environment, and individual specific components of variance for bone mineral density in 10- to 26-year-old females: A twin study. *Am J Epidemiol* 1998; **147**: 17–29

17 Snieder H, MacGregor AJ, Spector T. Genes control the cessation of a woman's reproductive life. A twin study of hysterectomy and age at menopause. *J Clin Endocrinol Metab* 1998; **83**: 1875–1880

11

SURVIVAL ANALYSIS
METHODS IN TWIN RESEARCH

Andrew Pickles

ABSTRACT

Analysis of censored or survival data can be problematic in twin studies. Censored phenotypic data occur when, although we don't know the actual value, we know that it is above (or below) some value. This commonly occurs with lifetime or age-of-onset data for those individuals who are still alive or yet to experience onset. It also occurs for psychological tests with floors or ceilings that restrict the range of measured variation and in biological assays that possess limits on the range of concentrations in which they are sensitive. There are perhaps three main approaches to the analysis of such data in biostatistics, and each of them have been adapted for use with twin data:

(i)   Frailty model extensions to proportional hazards models;
(ii)   Adaptations of ordinal response models;
(iii) Multilevel models for censored data.

In addition proposals have been made that extend the threshold liability model already current in biometrical genetics.

In this Chapter these models are described and contrasted, maximum likelihood and Bayesian methods for their estimation considered, and some further developments alluded to.

Age at onset has proved to be a useful marker of aetiological and genetic heterogeneity for a range of disorders, including prostate[1] and breast cancers [2] and Alzheimer's disease [3]. Family studies of depression have also suggested that, up to a point,[4] families of early onset probands have a higher liability. Thus data on age at onset are of considerable interest within genetics.

Biostatisticians refer to this class of data using the generic term 'survival data'. Survival data immediately conjures up images of data relating to time until death, but in fact is usually taken to include any data where the response is an age or time of onset or some latency time before a discrete response of some kind becomes evident. All my own experience of survival data is of this more general kind, with examples such as the age-of-onset of puberty, bike-pedalling of heart-patients until the onset of angina, and time until breakdown of foster placements. Survival data have two features that have tended to make the methods used to analyse them somewhat distinctive.

The first concerns the fact that time unfolds irreversibly, and particularly in the case of death, once the outcome has occurred individuals are essentially withdrawn from the sample under study, they being no longer at risk for that outcome. This has led to a natural focus of interest on an outcome probability measure for occurrence at time t conditional upon survival up to time t. Thus while standard regression analysis would focus on the distribution of t, in particular $f(t|x)$ adopting some convenient parametrization of the effect of some set of predictors x on the expected value of $f(t|x)$, much survival analysis focuses instead upon the conditional quantity $h(t|x)=f(t|x)/S(t|x)$, where $S(t|x)$ is the survival probability given by 1 minus the cumulative distribution function $(S(t|x)=[1-F(t|x)])$. This conditional quantity is referred to as the hazard function. Figure 11.1 illustrates the relationship between these quantities for the simple case of a time constant hazard $h(t|x)=0.5$; the hazard a horizontal line, the survival probability declining exponentially from 1 towards zero, and the density declining from 0.5 towards zero. The integrated hazard, which can be thought of as the total hazard to which an individual has been exposed so far, is also shown.

Whether modelling the density $f(t|x)$ or the hazard $h(t|x)$, the fact that both quantities cannot be negative, suggests the use of a log-transformation of the time-scale or a log-linear link. In either case the joint effect of risk factors is a multiplicative one. In the first case the effect on the expected failure time is one of a proportional increase or decrease, thereby giving rise to the name accelerated failure time model, since the effect is equivalent to time being 'speeded up' or 'slowed down'. In the second, the effect is a proportional increase or decrease in the hazard, thereby giving rise to the name proportional hazards model.

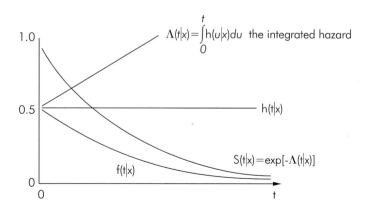

Figure 11.1 – Hazard, integrated hazard, density and survivor functions for a constant hazard model

The practical circumstances of having to wait until the response is observed almost inevitably results in some lifetimes or durations being observed incompletely, that is individuals being observed up to some point at which they are lost from the study, perhaps through data-collection being stopped, through attrition or through the subject becoming no longer eligible for the response under study. Such censoring of the response is not unique to survival data, also being common in psychometric test data with floors and ceilings, where a proportion of subjects are poorly discriminated by the test, either failing all the items or getting them all correct. Such responses can be modelled directly by means of the fitted distribution function. Censoring should be distinguished from sample truncation. In censoring we always know which observations have been censored since all observations are sampled. In sample truncation, for example where we know about the lifetimes of children only if they reached the age of a certain minimum age[5], the truncated individuals do not reach the sampling frame and we don't even know their frequency.[1]

PROPORTIONAL HAZARDS FRAILTY MODELS

Extending the standard proportional hazards model construction in which the effect of known measured explanatory variables was to multiply the risk intensity by a factor $\exp(x\beta)$, Clayton[6] suggested a multiplicative random effect or frailty z_i such that

$$\mu_i(t) = h\,(t\,|\,z_i,\,x_i) = z_i \exp(x_i\beta)\mu(t)$$

On the assumption of some appropriate distributional form or by the use of non-parametric mass-point representations for the frailty[5] and that for each twin within a pair the frailty is shared (i.e. $z_i = z_{ij}$) models of this form (see Figure 11.2) can be fitted to survival times for MZ and for DZ twins to yield zygosity specific estimates of the variance $\sigma^2(z)$. The greater the frailty, the greater the correlations in survival times and so the MD-DZ difference in the estimated frailty variances provides the basis for inference about possible genetic effects.

However, this model fails some basic conditions required for twin analysis. The classical twin model requires that the marginal MZ variance should equal the marginal DZ variance, since twins are assumed randomly sampled form the population of genes and environments. This os only true for the simple shared frailty model when the MZ and DZ frailty variances are the same, a circumstance implying the absence of genetic effects. Moreover, the single component of frailty confounds association between pairs with features of the marginal hazard function, such as non-proportionality in the covariate effects, that may not have been properly accounted for within the model. The fact that an estimate of $\sigma^2(z)$ can be obtained from data on singletons highlights this problem[7,8].

Hougaard[7,8] proposed a particular model construction that was not identifiable from singleton data alone. Subsequently, as illustrated in Figure 11.3, variously parametrized correlated frailty models were introduced to solve this problem. These involved both shared and unshared components of frailty with cross-

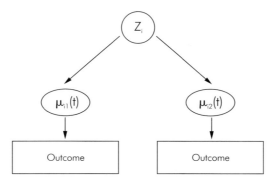

Fig. 11.2 – Simple shared frailty model

zygosity (total) variance constraints as required by the classical twin model. [9,10]

Pickles et al[9] presented three models that provided analytically tractable forms for the bivariate survivor function; one where the shared and unshared components were gamma distributed and combined additively on the hazard; one where they were positive stable law (PSL) distributed and combined additively, and the Hougaard[8] construction involving multiplicative effects with PSL distributions:

Additive Gamma

$$S(t_1,t_2) = [1 + \sigma_2^2 \Lambda(t_1)]^{(-1-\rho/\sigma_2^2)}[1 + \sigma_2^2 \Lambda(t_2)]^{(-1-\rho/\sigma_2^2)}[1 + \sigma_2^2\{\Lambda(t_1) + \Lambda(t_2)\}]^{(-\rho/\sigma_2^2)}$$

Additive Positive Stable Law

$$S(t_1,t_2) = \exp[-(1-\rho)\Lambda(t_1)^\alpha]\exp[-(1-\rho)\Lambda(t_2)^\alpha]\exp[-\rho\{\Lambda(t_1) + \Lambda(t_2)\}^\alpha]$$

Multiplicative Positive Stable Law

$$S(t_1,t_2) = \frac{\exp[-\{\Lambda(t_1)^{\alpha_1} + \Lambda(t_2)^{\alpha_1}\}^{\alpha_2}] \text{ for DZ twins}}{\exp[-\{\Lambda(t_1) + \Lambda(t_2)\alpha_1\}\alpha_2] \text{ for MZ twins}}$$

where $\Lambda(t)$ is the integrated hazard and the α's, ρ and σ^2 are frailty related parameters to be estimated.

For parametric baseline hazard functions estimation can be approached by direct maximum likelihood, and for the semi-parametric Cox model by E-M estimation[11,12] or Markov Chain Monte Carlo (MCMC) within a Bayesian framework.[13,14,15] The need for a flexible baseline hazard was illustrated in Pickles et al.[9] where it was shown how the estimates of genetic and environmental variance

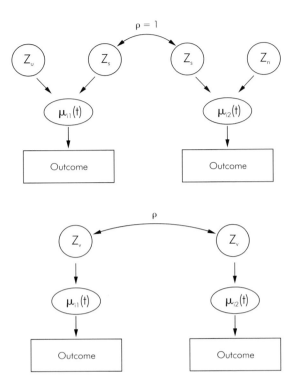

Fig. 11.3 – Correlated frailty: alternative parametrizations

were sensitive to getting the specification of the underlying hazard correct. A number of flexible parametric hazard functions exist, such as the piecewise exponential, and although the semi-parametric Cox model has considerable theoretical elegance, it often makes theoretical proofs difficult to obtain and may offer few practical advantages (see[11] p1481). The results in Pickles et al.[9] also suggest that although actual estimates of genetic and environmental effects differ according to the choice of distribution and the form of combination (additive or multiplicative), the values of test statistics for the presence of genetic effects are relatively stable. The extension of these models to include the segregation of a single gene of major effect has also been discussed.[14,16]

LINEAR MODEL ACCELERATED FAILURE TIME MODELS

Accelerated failure time (AFT) models have been generally overlooked but do have some advantages in the context of frailty.[17] The Weibull model is both a proportional hazards model and an AFT model. For a Weibull model with hazard function $\mu(t) = \gamma t^{v-1} \exp(\beta x)$ the AFT linear model representation is

$$\log(t_i) = -\frac{\log \gamma}{\nu} - \beta\frac{x_i}{\nu} + \frac{W_i}{\nu}$$

where W is a residual. The inclusion of a frailty term z within the hazard function results in an additional term, say U/ν, where U is -log(Z), and which is uncorrelated with the extreme value distributed W. This sort of model therefore becomes estimable by the more adaptable of the random effects/multilevel modelling software, though facilities need to be available for 'composite link functions' to allow for censored observations.

AFT models also provided an easy basis for generalizations to tackle the problem of

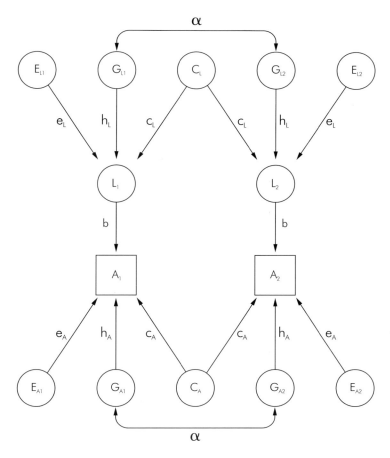

Fig.11. 4 – Path model for age-at-onset A, liability effects L where each is separately influenced by genetics, G., common environment C and specific environment E. Alpha is the coefficient of relationship. Drawn from Neale et al.[20].

measurement error in the recorded age-of-onset times.[18] This may be important in fields where reliance on retrospective reporting cannot be avoided or where only partial corroboration is available. It also generalizes easily to multivariate ages-of-onset, for example where onset is described by a set of onsets for individual symptoms or features. Pickles et al.,[19] consider such a model for the onset of breast development and menarche, finding evidence for these two onsets as being governed by a common set of genes even though the onsets are typically 18 months apart.

In all the models considered so far, correlation in age of onset and high familial liability are considered as both being due to shared frailty. Neale et al.[20] consider a more general model that combined an AFT type frailty model to model correlated age-of-onset, with an additional threshold-liability model.

Occurrence of disorder thus depended upon two dimensions each with genetic and environmental variation. The authors noted that very large samples would be required to distinguish convincingly dimension specific effects. Some caution might also be necessary to ensure that effects identified on the liability scale do not arise out of possible inappropriateness of the normality assumption for the age-of-onset distribution.

SIMPLE THRESHOLD MODELS FOR CENSORED ORDINAL RESPONSE

Distributional assumptions about the form of the age-of-onset distribution can be relaxed somewhat if the onset ages are grouped, and then modelled by means of a set of thresholds on some suitable distribution function, as shown in Figure 11.5.

For those familiar with the structural equation modelling program Mx[21] this is

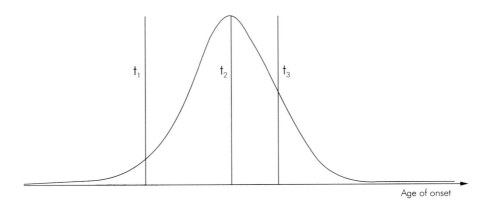

Fig. 11.5 – Age-at-onset categories in an ordinal response model

perhaps the most straightforward approach to the analysis of twin survival data. Pickles et al.,[22] describe the layout of the data into simple frequency tables of age-of-onsets for twin 1 by twin 2, though the Mx setup described there requires some very slight modification to conform with new Mx enhancements. However, in practice, the presence of censored durations requires the tables to be disaggregated into groups with similar censoring times (and usually common censoring times for both twins). For some datasets this requires a large number of frequency tables to be input in order to achieve an efficient analysis. As was the case in the AFT models, some allowance can be made for errors in the reporting of age-of-onset.

Table 11.1 gives results for an analysis of the age-of-onset of a male pubertal feature, maternal reports of 'voice breaking' in 95 MZ and 85 DZ pairs of twin sons. Estimates of thresholds like those shown in Figure 11.5 are presented. The variance components estimates indicate striking genetic variance. Also illustrated is the impact of measurement error in the dates of onset as recalled by mothers, measurement error that was correlated across twins. Without correction, these errors result in an inflated estimate of the common environment effect.

GENE-ENVIRONMENT INTERACTION

In all of the approaches outlined above, extension to allow for interactions between genes and known environmental exposures is relatively straightforward. Using

Table 11.1 – Parameter Estimates and Relative Fit: 6 Age-Group Model (11 and less to 16 or more)

	Genes & Environment	Genes, Environment & recall error correction
Standardized Variances		
Genetic	0.85	0.91
Common Environment	0.10	0.00
Specific Environment	0.05	0.09
Thresholds		
1	−2.41	
2	−1.76	
3	−0.99	
4	−0.34	
5	0.45	
6	1.08	
Chi-square	131.49	99.28
D.f.	257	252

these methods, there is increasing interest in the pharmaceutical industry at the possibility of exploring response heterogeneity within genetically informative designs.

DYNAMIC FRAILTY

That individuals should possess time constant frailties can at best be considered a first approximation. It is to be expected that the unmeasured environments of individuals change with age, that the level of expression of different genes changes with age and that the effect of the same level of gene expression changes with changing environmental exposure. Although much remains to be done, elements of some possible model constructions are available (e.g.[23,24]).

AUXILIARY VARIABLES

In most circumstances there is considerable pressure to extract information about effects in survival studies before sufficient failures (e.g. deaths) have occurred that would provide the analysis with much power. However, in some circumstances biological markers may exist that reflect the development of disease and that can be measured longitudinally prior to death. For example, the CD4 count of patients in an AIDS clinical trial might be measured. Useful efficiency gains may be made by exploiting such measures of morbidity to support the information on mortality[25,26]. The simultaneous modelling of gold-standard and auxiliary information might follow the trend in behaviour genetics towards fully parametric multivariate modelling, or might be undertaken using developments of the more cautious semi-parametric approach more commonly preferred in epidemiology[27]. In either case, the use of this auxiliary information for more efficient sampling designs opens up new empirical opportunities.

CONCLUSIONS

This brief review has covered just some of the range of distinctive and varied models and estimation methods available (but see[28], for another example). These allow the fitting of genetically informed models to survival and age-of-onset data from twins and families. Nonetheless much work remains to be done, particularly in relation to their construction around biologically plausible models of the process of development and pathogenesis, and in the intelligent combination of survival data with auxiliary information.

ACKNOWLEDGEMENT

This work was partially supported by grants MH-45268 from the US NIMH.

REFERENCES

1   Carter, B.S., Beaty, H.B., Steinberg, G.D., Childs, B. and Walsh, P.C. (1992) Mendelian inheritance of familial prostate cancer. *Proceedings of the National Academy of Sciences U.S.A.* 89, 3367–3371.

2   Claus, E.B., Risch, N. and Thompson, W.D. (1990) Using age at onset to distinguish between subforms of breast cancer. *Annals of Human Genetics* 54, 169–177.

3   Chartier-Harlin, M.C., Crawford, F., Houlden, H., et al. (1991) Early-onset Alzheimer's disease caused by mutations at codon 717 of the beta-amyloid precursor protein gene. *Nature* 353, 844–846.

4   Harrington R., Rutter M., Weissman M., et al. (1997) Psychiatric disorders in the relatives of depressed probands. I. Comparison of prepubertal, adolescent and early adult onset cases. *Journal of Affective Disorders* 42, 9–22

5   Hougaard, P., Harvald, B. and Holm, N.V. (1992) Measuring the similarities between lifetimes of adult Danish twins born 1881–1930. *Journal of the American Statistical Association* 87, 17–24.

6   Clayton, D.G. (1978) A model for association in bivariate life-tables and its application in epidemiological studies of familial tendency in chronic disease. *Biometrika* 65, 141–151.

7   Hougaard, P (1986a) Survival models for heterogeneous populations derived from stable distributions. *Biometrika*, 73, 387–396.

8   Hougaard, P. (1986b) A class of multivariate failure time distributions. *Biometrika* 73, 671–678.

9   Pickles A., Crouchley R., Simonoff E., et al. (1994) Survival models for developmental genetic data: age of onset of puberty and antisocial behaviour in twins. *Genetic Epidemiology* 11, 155–170.

10 Yashin, A.I. and Iachine, I.A. (1995) Genetic analysis of Durations: correlated frailty applied to survival of Danish twins. *Genetic Epidemiology* 12, 529–538.

11 Anderson, P.K., Klein, J.P., Knudsen, K.M. and Tabanera y Palacios, R. (1997) Estimation of the variance in Cox's regression model with shared gamma frailties. *Biometrics* 53, 1475–1484.

12 Murphy, S.A. (1994) Consistency in a proportional hazards model incorporating a random effect. *Annals of Statistics* 22, 712–731.

13 Thomas, D.C., Langholz, B., Mack, W and Floderus, B. (1990) Bivariate survival models for analysis of genetic and environmental effects in twins. *Genetic Epidemiology* 7, 121–135.

14 Thomas, D.C. and Gauderman, W.J. (1996) Gibbs sampling methods in genetics. In Gilks, W.R., Richardson, S. and Spiegelhalter, D.J. (Eds.) *Markov Chain Monte Carlo in Practice.* Chapman and Hall, London. Pp419–440.

15 Clayton, D.G. (1991) A Monte Carlo method for Bayesian inference in frailty models. *Biometrics* 47, 467- 485.

16 Hogzhe L. & Thompson, E. (1997) Semiparametric estimation of major gene and family-specific random effects for age of onset. *Biometrics* 53, 282–293.

17 Keiding, N., Andersen, P.K. and Klein, J.P. (1997) The role of frailty models and accelerated failure time models in describing heterogeneity due to omitted covariates. *Statistics in Medicine* 16, 215–224.

18 Pickles A., Pickering K. and Taylor C. (1996) Reconciling recalled dates of developmental milestones, events and transitions: A mixed GLM with random mean and variance functions. *Journal of the Royal Statistical Society* Series A. 159, 225–234.

19 Pickles, A., Pickering, K., Simonoff, E., Meyer, J., Silberg, J. and Maes, H. (1998) Genetic 'clocks' and 'soft' events: A twin model for pubertal development and other recalled sequences of developmental milestones. *Behavior Genetics*. 28, 243–253.

20 Neale, M.C., Eaves, L.J., Hewitt, J.K., Maclean, C.J., Meyer, J.M. and Kendler, K.S. (1989) Analysing the relationship between age at onset and risk to relatives. *American Journal of Human Genetics* 45, 226–239.

21 Neale, M.C. (1996) *Mx Manual*. Medical College of Virginia. Richmond, VA, USA.

22 Pickles A., Neale M., Simonoff E., et al. (1994) A simple method for censored age of onset data subject to recall bias: Mother's reports of age of puberty in male twins. *Behavior Genetics* 24, 457–468.

23 Yue, H. and Chan, K.S. (1997) A dynamic frailty model for multivariate survival data. *Biometrics* 53, 785–793.

24 Self, S. (1993) A regression model for counting processes with a time-dependent frailty. Mimeo.

25 Hogan, J.W. and Laird, N.M. (1998) Increasing efficiency from censored survival data. *Statistical Methods in Medical Research* 7, 28–48.

26 Fleming, T.R., Prentice, R.L., Pepe, M.S., Glidden, D. (1994) Surrogate and auxiliary endpoints in clinical trials, with potential applications in cancer and AIDS research. *Statistics in Medicine* 13, 955–968.

27 Clayton, D.G., Spiegelhalter, D., Dunn, G. and Pickles, A. (1998) Analysis of binary longitudinal data from multiphase samples (with discussion). *Journal of the Royal Statistical Society, Series B*. 60, 71–102.

28 Meyer, J.M. and Eaves, L.J. (1988) Estimating genetic parameters of survival distributions: a multifactorial model. *Genetic Epidemiology* 5, 265–275.

12

GENE-ENVIRONMENT INTERACTION AND TWIN STUDIES

Nick Martin

ABSTRACT

Gene-environment interaction (GxE) refers to the fact that different geno-types respond differently to the same environment. GxE is related to the statistical concept of heteroscedasticity, implying heterogeneity of variance. Monozygotic (MZ) twins can be used to detect systematic heteroscedasticity, but it is important first to define the scale of interest. For a continuous trait (eg. bone density or blood pressure) a non-normal distribution will generate scale-dependent GxE which can be removed by appropriate transformation. For a dichotomous trait (eg. disease status) non-additivity on the risk scale can occur even when there is additivity on the liability scale that underlies it, and the reverse applies too. MZ twins provide a unique opportunity to detect various types of GxE. For measured genotypes, MZ intrapair variances should be compared to detect genes affecting environmental sensitivity. Such "variability" genes are potentially even more important than "level" genes and provide a sound rationale for collecting DNA from MZ as well as DZ twins. The power of such studies is examined. An example is given of using MZ twins to detect a variability gene which predicted response to an environmental intervention in a clinical trial.

"No aspect of human behaviour genetics has caused more confusion and generated more obscurantism than the analysis and interpretation of the various types of non-additivity and non-independence of gene and environmental action and interaction". Unfortunately, this statement is as true today as when it was written 21 years ago by Eaves et al.[1] In that paper, Eaves and colleagues outlined many of the issues and misconceptions surrounding genotype × environment interaction (GxE) and for a full treatment readers should consult it. Here I shall simply draw attention to some of the key issues as they relate to the design and analysis of twin studies.

DEFINITIONS

First of all we should make quite clear what we mean. Many people who should know better use the term GxE to denote that both genes and environment are important. Apart from the triviality of such a sentiment (what manifestation of life is not ultimately coded for by genes, and what life is not dependent on the supply of air, water, food from the environment ?), a better term is "genotype-environment co-action". We should reserve for the term GxE its statistical sense of different genotypes responding differently to the same environment; or viewed from the other end, some genotypes being more sensitive to changes in the environment than others (different reaction ranges).

Once put this way, it is immediately clear that GxE must be related to the well known statistical concept of heteroscedasticity – that a single variance is not adequate to describe the whole population i.e. certain subpopulations are more variable than others. This heterogeneity of variance may follow no particular pattern (unsystematic) or there may be some simple relation between the mean of subgroups and their variance. For example, measurements of complex processes frequently follow a lognormal distribution in which the measurement error is proportional to the mean value at different parts of the distribution. Analysis of variance leading to statements that so much of the variance is due to main effects and so much due to error are then clearly meaningless since, these proportions will depend on where one lies on the scale of measurement. Fortunately, such systematic mean-variance relationships are easily removed by an appropriate transformation of the scale of measurement (in this case, logarithmic).

INFERRING G × E FROM TWIN DATA

In fact, MZ twins offer a unique opportunity to detect such systematic heteroscedasticity, which manifests itself, formally, as one form of GxE, a fact brilliantly realised in their ground-breaking paper by Jinks & Fulker in 1970.[2] They pointed out that the *difference* between MZ twins is an estimate of the environmental effect, whereas the *sum* of MZ twins' scores is a function of their (shared) genetic deviation and their family environmental deviation. Assuming that there is some genetic contribution to the trait – as evidenced by a greater MZ than DZ correlation – any relationship between the absolute differences of MZ pairs and their corresponding sums is evidence of GxE interaction. Jinks and Fulker[2] suggested that as a preliminary step to any biometrical genetic analysis of twin data, one should therefore regress the MZ absolute pair differences on their pair sums to check for GxE.

In practice, countless applications of the Jinks and Fulker[2] GxE test over the ensuing 28 years have shown that the GxE detected is completely predictable as a function of the nonnormality of the scale distribution and that it can be removed by an appropriate transformation of the scale of measurement. Thus, positive skewness (tail to the right) produces positive correlations between MZ absolute differences and sums, and negative skewness produces a negative correlation. The oft-observed "basement-ceiling" effect in psychometric scales, where there is good discrimination of differences in the middle of the range but a bunching at the low and high extremities, produces an inverse U-shaped quadratic relationship of means and intrapair differences. These relationships are illustrated in Figure 12.1.

This notion, that the appearance of GxE depends on the scale of measurement becomes even more critical when we move from continuous variables to categorical, of which the most common instance is the dichotomous case of disease status – affected versus unaffected. The demonstration of an interaction between genotype and environment on disease risk may suggest dramatic public health messages of the type we all say Amen to in our grant applications. But it may not be the most revealing way to view etiology. Perfect additivity of genetic and environmental influences on the normal liability scale can translate to non-additive GxE interactions on the scale of disease risk. Only in the case of cross-over interactions which occur when certain genotypes have extreme sensitivity to environmental changes whilst others are relatively impervious, is one guaranteed to observe GxE on both the liability and risk scales[3]

G × E USING GENOTYPE DATA

The unique value of MZ twins in detecting biometric GxE can also be exploited to search for interactions between particular measured (ie. typed) genes and the environment. Magnus et al[4] showed that the intrapair variance for cholesterol in 22 MZ pairs who were blood group M- (ie. blood group N) was significantly greater than in 75 pairs who were M+ (ie. blood groups M and MN). From this observation Berg et al[5] introduced the idea of "variability genes" as opposed to "level genes". Level genes affect the mean expression of a trait, or prevalence in the case of a disease, and are the usual target of association studies. However, it is possible that there are genes that have no effect on the mean expression level but have a greater or lesser variance of expression. It is not difficult to think of molecular mechanisms – eg. promoters of different binding efficiency, to explain the existence of such variability genes. Neither are they a novel concept to plant and animal breeders, who have long been familiar with the existence of genes influencing environmental sensitivity. Indeed, much of the increase in crop yields this century has been due to selection for genes able to respond to increased fertiliser doses, as opposed to genes with higher yields in the average environment.

The potential importance of such variability genes to human health has yet to dawn upon the gene-hunting community. Potentially, such genes are even more important than level genes. An allele which increases environmental variance by 50% (the approximate effect size of the N allele on cholesterol variability) will increase the proportion of cases above the second standard deviation by more than two fold, and those above the third SD by more than five fold. Unfortunately,

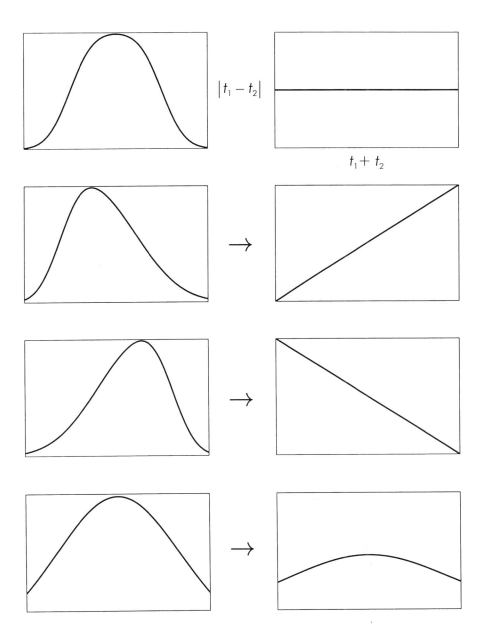

Figure 12.1 – Relationship between scale of measurement and the Jinks & Fulker G × E test: regression of MZ pair absolute differences on pair sums.

however, the power of the variance ratio test to detect heterogeneity of variances is not large. For the favorable case of the MN blood group with two alleles of high frequency, around 530 MZ twins would need to be typed for 80% power at the 5% significance level.[6] A forward-looking (and rich!) gene-hunting company would be obtaining good phenotypes on large numbers of MZ twins against the day (soon promised) when it will be a cheap option to scan the genome for tens of thousands of closely spaced SNP's, any of which may be (in linkage disequilibrium with) such a variability gene.

To demonstrate the importance of such variability genes, we predicted from our findings on intrapair variances of MZ twins that serum lipid levels of individuals who were blood group N would respond more to a low fat diet than those who were M+. We had our chance to test our hypothesis in the course of a case-control study to investigate the effect of diet on the recurrence of colorectal polyps.[7] One of the dietary factors was an intervention to reduce fat in the diet, and over an 18 month period we indeed observed that individuals who were blood group N had the greatest lowering of their LDL cholesterol. However, the most significant effect was the contrast between the two homozygotes (blood groups M and N) and the heterozygote MN group who appeared most impervious to the changed diet. (Figure 12.2) This is in line with a large body of population genetic theory which suggests that heterozygotes are most buffered against environmental extremes. In the modern world, with ready access to fatty diets, MN heterozygotes may not be at such risk of succumbing to hypercholesterolaemia. On the other hand, people of blood group N may be the best targets of cholesterol-lowering therapies. In general, the identification of such variability genes is likely to have great potential in targetting public health measures not just to the most susceptible individuals (the usual diagnostic sales pitch) but also to the individuals most responsive to treatment.

There are some genes whose alleles appear to affect both level and variability, and here the choice of scale is very important. We investigated the effect of isozyme alleles at the protease inhibitor (PI) locus on chromosome 14, on levels of alpha-1-antitrypsin concentration (AAT) and its activity measured as elastase inhibitory capacity (EIC).[8] The most common alleles are M1, M2 and M3 and we found that the genotypic means for the various combinations of these alleles were heterogeneous. For AAT we found that the variance of homozygotes for these alleles was significantly greater than for their heterozygotes (echoes of the MN blood group/cholesterol story above), and that this applied whether we used the raw scale or a logarithmic transformation. For EIC there was also hetero-geneity of variance, but its direction depended on the scale of measurement, suggesting that it might be possible to find a transformation for homoscedas-ticity. In this case, it is of prime importance to decide the scale of most clinical relevance ahead of time, before deciding that one has, or does not have GxE to worry about.

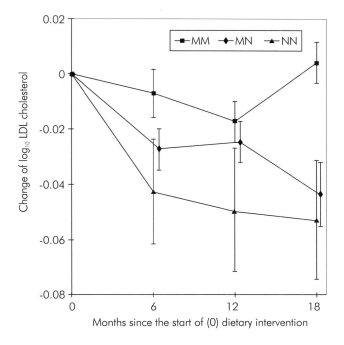

Figure 12.2 – Changes of LDL (log10mm. L⁻¹) cholesterol at 6, 12 and 18 months together with standard errors of the change scores. Genotypes are represented as different symbols as in the legend.

CONCLUSIONS

The power of MZ twins to detect susceptibility genes leads one to a more general consideration of the power of twin studies to detect GxE of various types. In a series of simulations, we considered measured genotypes, G_m, (i.e. typed polymorphisms), measured environmental influences, E_m (such as dietary fat intake) and their residual counterparts G_r and E_r which can be estimated from the classical twin study. In fact it is possible to estimate all six two-way interactions between these four terms, except for $G_r \times E_r$ which is totally confounded with E_r. Analysis of as few as 600 MZ and 600 DZ twins have been shown to have the power to detect quite appreciable GxE effects of considerable practical importance.[9] In the end, our ability to detect such effects may not depend so much on the quality and quantity of the genotypic and environmental measures at our disposal as on our skill, time and energy (i.e. money!) to perform the appropriate analyses.

REFERENCES

1 Eaves LJ, Last KA, Martin NG, Jinks JL. A progressive approach to non-additivity and genotype-environmental covariance in the analysis of human differences. *British Journal of Mathematical and Statistical Psychology* 1977; **30**: 1–42

2 Jinks JL, Fulker DW. Comparison of the biometrical genetical, MAVA and classical approaches to the analysis of human behavior. *Psychological Bulletin* 1970; **73**: 311–349

3 Kendler KS, Eaves LJ. Models for the joint effect of genotype and environment on liability to psychiatric illness. *American Journal of Psychiatry* 1986; **143**: 279–289

4 Magnus P, Berg K, Borreson A-L, Nance WE. Apparent influence of marker genotypes on variation in serum cholesterol in monozygotic twins. *Clinical Genetics* 1981; **19**: 67–70

5 Berg K, Kondo I, Drayna D, Lawn R. "Variability gene" effect of cholesteryl ester transfer protein (CETP) genes. *Clinical Genetics* 1989; **35**: 437–445

6 Martin NG, Rowell DM, Whitfield JB. Do the MN and Jk systems influence environmental variability in serum lipid levels ? *Clinical Genetics* 1983; **24**: 1–14

7 Birley A, MacLennan R, Wahlqvist M, Gerns L, Pangan T, Martin NG. MN blood group affects response of serum LDL cholesterol level to a low fat diet. *Clinical Genetics* 1997; **51**: 291–295

8 Martin NG, Clark P, Ofulue AF, Eaves LJ, Corey LA, Nance WE. Does the PI polymorphism alone control alpha-1-antitrypsin expression? *American Journal of Human Genetics* 1987(a); **40**: 267–277

9 Martin NG, Eaves LJ, Heath AC. Prospects for detecting genotype × environment interactions in twins with breast cancer. *Acta Geneticae Medicae et Gemellologiae* 1987(b); **36**: 5–20

13

WHY 'COMMON ENVIRONMENTAL EFFECTS' ARE SO UNCOMMON IN THE LITERATURE

John L. Hopper

ABSTRACT

Familial aggregation is a feature of almost all human traits and diseases. Twin studies are important because they have the potential to identify traits for which genetic factors may explain at least a portion of variation. However, under the classic twin model it is assumed that the effects of environmental factors shared by twins (also often referred to as 'common environment' effects) are independent of zygosity. Consequently, the possibility that the trait-specific environmental effects could be more correlated within MZ pairs than within DZ pairs is, in effect, completely dismissed by researchers who rely solely on this model. Furthermore, under the classic twin model there is usually little statistical power to detect shared environment effects, and this often leads to a parsimonious model that attributes *all* within-pair aggregation to genetic factors. This 'finding' often appears to be incredulous, especially to researchers of the non-genetic factors involved in the trait. Therefore, the typical lack of evidence for shared or common environment effects resulting from textbook application of classic biometrical modelling may not be a proper interpretation of reality. Such statistical analyses, biased towards a genetic explanation and then interpreted as evidence (if not proof!) that genetic factors alone are causing familial aggregation, may be misleading. These and a number of other related issues are illustrated with reference to particular data sets and analyses. It is suggested that twin researchers might benefit from taking a more critical approach to model fitting, and in particular, trying to falsify genetic hypotheses.

Familial aggregation is important. It is a feature of almost all human traits and diseases, and even modest clustering for a binary trait (such as a disease state; affected or unaffected) cannot occur without there existing one or more strong underlying familial risk factors.[1-3] The conundrum of whether family clustering is due to genetic inheritance, cultural transmission, shared non-genetic factors or some combination, or even interaction, of the above, is not easily resolved. Informative family designs, such as those based on twins, adoptees or migrant families, are potentially useful, but only if the design and statistical analysis can allow discrimination between the multiplicity of possible explanations for familial aggregation. Statistical models that are biased towards a genetic explanation, and then interpreted as evidence (if not proof!) that genetic factors alone are causing familial aggregation, may be misleading.

The basic features of biometric modelling date back to the seminal paper by R.A. Fisher, published in 1918.[4] Sewall Wright's path analysis models, published shortly thereafter,[5] provide a diagrammatic way of picturing the sources and effects of different causes of variation. Over the last few decades the statistical methods for analysing twin and family data, and the computation of estimates and measures of imprecision and lack of model fit, have been formalised using maximum likelihood theory.[6–9] More sophisticated statistical methods and computational power are now available, but in application these advantages have not necessarily been exploited or understood.[10]

The classic twin model relies totally on the assumption that the effects of environmental factors shared by twins (also often called 'common environment effects') are independent of zygosity. In practice, this often results in an estimate of the 'shared environment effect' that is not statistically significant. But the classic twin model generally has little statistical power to detect such shared environment effects, so even substantial effects may not be detected. This then leads to the 'parsimonious model', the model that explains as much variation as possible with the fewest parameters, attributing *all* within-pair aggregation to genetic factors. This 'finding' often appears incredulous, especially to researchers of the non-genetic factors involved in the trait. The typical lack of evidence for shared or common environment effects resulting from textbook application of classic biometrical modelling may not be a proper interpretation of reality.

Why is this important? The last few decades of the 20th century has seen the rapid emergence of molecular and statistical methods for scanning the genome to identify regions, if not specific loci, involved with expression of a phenotype of interest. The new millennium promises that this process will become easier, especially once the human genome has been mapped. Regions spanning particular 'candidate genes' can be targeted and tested for allelic association (or linkage) with the trait. This endeavour will not be cheap, however. In order to produce convincing evidence, large numbers of different types of individuals and relatives will need to be phenotyped, and a large number of loci genotyped. Cost efficiencies will need to be sought, and one obvious way is to first choose traits for which there is good evidence that a substantial proportion of variation is under genetic control. It will be wasteful of time and resources if geneticists are misled into wild goose chases by simplistic biometric analysis which over-estimate genetic variation through having ignored, or addressed poorly, the role of non-genetic factors in causing trait variation. In order to understand what is involved, a number of issues need to be taken into account. Some of these are discussed below, and illustrated with reference to particular data sets and analyses.

SOME ISSUES

There is an inherent bias towards 'finding genetic effects' when twin data is analysed using the classic twin model. There are two major reasons for this. The first comes about because any excess in MZ trait correlation over DZ trait correlation is attributed to 'genes', effectively ignoring any implications of the obvious – that genetically-identical, monozygotic (MZ) twins lead more similar lives than do fraternal, dizygotic (DZ) twins. The second is a consequence of the usual lack of statistical power to detect shared environment effects under the assumptions of the classic twin model. This modelling paradigm is geared towards concluding that a genetic model is the more 'parsimonious' description of the twin correlations, even though there may be quite substantial non-genetic factors shared by twins.

A statistical model is a convenient mathematical representation that might be useful for making predictions, and guiding scientific endeavours. It is of necessity simplistic, and is usually based on strong assumptions that need to be tested in each context in which it is applied. Its utility, therefore, is critically dependent on the validity of those assumptions. The Classic Twin Model is no exception.

1. The Underlying Assumption of the Classic Twin Method

Under the classic twin model, and for any ordinal trait, if the correlation between MZ twin pairs, r_{MZ}, is greater than for DZ twin pairs, r_{DZ}, the data is said to be consistent with, but importantly does not prove that, genetic factors explain a proportion of variation. (The caveat is often ignored in application.) This interpretation is premised on the key assumption that the sum of all trait-specific environmental *effects* shared by twins within a pair, referred to as 'common' or 'shared' environment effects, are independent of zygosity. That is, the classic twin method is valid only if the effects shared within twin pairs are of the same magnitude in MZ pairs as they are in DZ pairs. (Often this assumption is worded in terms of the extent of sharing the environment, whereas it is really about the correlation in effect size – for the trait in question – between twins within a pair.)

Under this assumption, one can model the effects of shared genes and of shared environments together, and this is most conveniently done by assuming that those two effects are statistically independent. The modelling and interpretation becomes more complicated if these simplistic assumptions do not hold, and in practice one cannot rule out gene-environment and gene-gene interactions and correlations, and non-additive genetic and environmental effects For clarity, however, the following discussion will revolve around the usual assumptions of independent and additive genetic and environmental effects.

The major problem is that, under the classic twin model, the die is loaded in favour of a genetic explanation. This is because *any* excess in MZ correlation compared to DZ correlation is interpreted as being solely due to genes. The possibility that environmental effects could be more correlated within MZ pairs that within DZ pairs is, in effect, completely dismissed by twin researchers who rely solely on the classic twin model.

If the assumption above is not correct, in that the environment effects shared within MZ pairs are greater than within DZ pairs, then one can never be sure that finding $r_{MZ} > r_{DZ}$ means anything in terms of 'proof for the existence of genetic causes of variation'! Rather than just blandly accepting or not accepting this assumption, a quantitative scientist should try to assess the extent to which the assumption is valid. Note that this needs to be considered anew for each trait, because the assumption refers to the common environment effects particular to the trait being studied; see 3. below.

When challenged, researchers beholden to the classic twin model will spiritedly defend the assumption, but rarely with data specific to the trait of interest. This may in part be because the more commonly used methods and statistical packages, such as those based on structural modelling[9], do not easily lend themselves to a flexible analysis. On the other hand the multivariate normal model allows much greater flexibility, as demonstrated by a number of examples below that find evidence supporting the existence of shared environmental effects. It requires a greater sophistication in mathematical and programming skills.

2. Statistical power

Let us presume, for now, that the above critical assumption of the classic twin model actually does hold, and that there are no sources of non-additive genetic variance. The estimate of the proportion of variance attributed to common or shared environment effects is given by $C = 2r_{DZ} - r_{MZ}$. For simplicity, let the numbers of twin pairs of each zygosity be equal; i.e. $n_{MZ} = n_{DZ} = n$. Suppose the true values of r_{MZ} and r_{DZ} are both <0.5, so their standard errors can be approximated by $(n-3)^{-1/2}$, and that their estimates are independent of one another. The standard error of C is then approximately $[5/(n-3)]^{1/2}$. To have 80% power at p = 0.05 (one-sided) to detect a common environment effect that explains c% of variance or more, n will need to be large enough for the standard error to be greater than 2.5c. (The argument applies if higher values of r_{MZ} and r_{DZ} are involved, but approximate computation of the standard error is simpler for this typically observed range.)

Table 13.1 shows the number of pairs required for different values of c. If n is less than 100, common environment effects will be difficult to detect even if they explain 50% of variation. Even with samples of 1,000 pairs of each zygosity, substantial common environment components of variance will more than likely go undetected by the usual modelling approach.[11]

As an example, suppose the true state of affairs is that additive genetic factors explain 40% of the variance, and environmental and other non-genetic effects that are independent within twin pairs explain 40% of the variance. This leaves 20% of

Table 13.1 – The number of twin pairs, n, necessary for there to be 80% power at the 0.05 level of significance to detect a common environment effect that explains c% of variance or more.

n	c
<34	100
100	50
500	25
>3,000	10

the variance explained by environmental factors whose effects are correlated within a pair. It will also be assumed that the assumption of the classic twin model holds, so the above correlation in shared environment effects is the same for MZ pairs as it is for DZ pairs. In this case, the expected trait correlations within pairs are r_{MZ} = 0.6 for MZ pairs, and r_{DZ} = 0.4 for DZ pairs. Suppose 100 MZ and 100 DZ pairs are studied. Then the expected value of C is 0.2, and its standard error is about 0.23. Consequently, the probability that a value of C is observed that would not be considered 'significant' would be about 0.2.

According to the usual modelling paradigm, should the test for a significant shared environment effect fail to achieve nominal significance, it is presumed that there is *no* other source of within-twin pair correlation than genetic factors! The 'parsimonious' model will then only allow for additive genetic factors and independent environmental factors. For the above situation, the maximum likelihood estimator of the proportion of variation due to genes, A, will on average be about 0.6. That is, in 1 in 5 instances the estimate of genetic contribution to variance will be overestimated by about 50%.

3. The assumption of the classic twin model needs to be considered anew for each application

The assumption of the classic twin model, that the extent to which twin pairs share the environmental *effects* is independent of their zygosity, is trait specific, and must be considered anew for each trait, in each environment. That is, one cannot rely on an overall statement that this assumption is valid, even for a specific trait. It must be tested in the context in which it is being applied.

4. Testing the Classic Twin Model Assumption

How might one go about testing that the within-twin pair shared environment effects are independent of zygosity? One way is to measure whether pairs of relatives live together, and if they live apart, how long since they cohabited, and see if the trait correlation between relative pairs varies plausibly. For example, does this correlation increase the longer they have been cohabiting, and does it decrease the longer they have lived apart? This approach was demonstrated for blood lead levels measured in non-twin nuclear families,[8,12] where evidence suggesting a substantial effect of shared sibling environment that abated gradually over adult life was observed and quantified.

Inclusion of data from non-twin siblings can help. Comparison of the DZ correlation with the same-sex non-twin sibling correlation allows estimation of DZ-specific common environment effects. For example, when data from a twin family study carried out in the UK were analysed along these lines,[13] evidence for effects of a common twin environment were estimated, separately, for alcohol consumption, anxiety and depression. For alcohol consumption there was no evidence that these effects were zygosity dependent, although the effects did differ depending on whether twins lived together or apart. On the other hand, for both anxiety and depression, the estimated effects of common twin environment appeared to be qualitatively and quantitatively different between MZ and DZ pairs.

Application to measures from the 16-PF questionnaire, this time measured in nuclear families, gave further interesting findings.[14] For example, for a brief measure of IQ there was evidence for small cohabitation effects within the family, except for father-offspring pairs. For the trait 'sizia versus affectia' (out-going versus reserved), the effect of cohabitation for mother-daughter pairs was estimated to be negative, in that it resulted in a lower correlation. This highlights the fact that cohabitation can, for some traits, result in relatives becoming *less* similar, perhaps due to what has been called 'competition' effects.

5. How well do the data fit the predictions of a genetic model?

It is one thing to say that a certain model is the 'best' fit, or the most parsimonious (in that extra parameters do not appear to improve the fit more than expected by chance alone), but that doesn't mean that the fit is 'good'. Scientists need to rigourously test their model if they are to be convinced of its utility.

The original model derived by R.A. Fisher more than 80 years ago[4] can be used to do more than just estimate variance components; it predicts that if additive genetic factors are the only genetic effects involved, r_{MZ} will be precisely twice r_{DZ}. Using the delta method,[15] and replacing correlations by their observed values, Var (r_{MZ}/r_{DZ}) is approximately $Var(r_{MZ})/r_{DZ}^2 + r_{MZ}^2 Var(r_{DZ})/r_{DZ}^4$. Again, for simplicity, let $n_{MZ} = n_{DZ} = n$, and consider the case where both r_{MZ} and r_{DZ} are less than 0.5, so that their variances are each approximately $(n-3)^{-1}$. Then under the null hypothesis that the correlation in MZ pairs is twice the correlation in DZ pairs, the variance of the ratio of correlation estimates is approximately $5/[n(r_{DZ})^2]$. That is, the precision of the estimate of the ratio of correlations rapidly increases as r_{DZ} gets closer to zero. If the true values are $r_{MZ} = 0.4$ and $r_{DZ} = 0.2$, Var (r_{MZ}/r_{DZ}) is approximately $125/n$. In this case, r_{MZ}/r_{DZ} has an expected value of 2, and a standard error of more than 1 if n is less than 100, and about 0.5 if n = 500. Even if 1,000 twin pairs are studied, reasonably large deviations from the 2:1 ratio of MZ:DZ correlations can occur; about 1 in 20 times the ratio will be greater than 2.6 or less than 1.4. Therefore, very large samples are needed in order to be able to assess critically the genetic model. In practice, this is rarely if ever attempted.

6. Do the observed correlations between relatives adhere closely to those predicted by a fitted genetic model?

Often data from different types of relatives, including twin pairs, are analysed together. A good test of a model is to examine the consistency of predicted versus observed trait correlations for the different relative groups.

For example, in the 1918 paper referred to above,[4] Fisher used data on stature (height) in pairs of relatives collected by Francis Galton, to demonstrate his biometric model. In a recently conducted study of cardiovascular related variables in 783 families, including 89 MZ and 86 DZ twin pairs, of adults living in Melbourne, Australia (total sample size almost 3,000 individuals),[16] the correlations for age- and sex-adjusted height adhered closely to the pattern anticipated under a model which attributes all familial aggregation to additive genetic factors, and includes assortative mating. Spouse pairs had a correlation of just under 0.3. Parent-offspring pairs had a correlation of 0.46, with mother-child and father-child correlations being 0.50 and 0.44 and not significantly different. These parent-offspring correlations also varied little by sex of the child. Moreover, sibling correlations were similar to parent-offspring correlations, being 0.45 overall and having a tight range from 0.42 in sister-sister pairs, to 0.44 in brother-brother pairs, and 0.50 in brother-sister pairs. These small discrepancies were consistent with chance. For DZ pairs the correlation was 0.44, and no different from the correlation in other types of siblings. That is, correlations for all types of first-degree relative pairs were strikingly similar, in accordance with the prediction of an additive genetic model. For MZ pairs, the correlation was 0.74, significantly greater than for first-degree relatives, and consistent with an additive model once assortative mating was taken into account, as in Fisher's original paper.[4]

In contrast, for body mass index (BMI) the pattern of correlations increased progressively from spouse pairs (0.24) to parent-offspring pairs (0.26) to sibling pairs (0.33) to DZ pairs (0.56) to MZ pairs (0.69). Although such a pattern is not inconsistent with an additive model, it is consistent with the anticipated effects of factors shared by family members, such as food intake, physical activity and other lifestyle variables related to weight. These factors will typically be shared most closely by MZ pairs, then DZ pairs. Siblings are likely to share them less the greater the difference in their ages. Parents and their offspring are also likely to share them less again, especially if the children – or parent – are spending considerable amounts of time away from the family home. The postulated gradation of shared environmental effects above can be tested empirically, provided relevant information is recorded.

For the cardiovascular-related measures diastolic blood pressure, HDL cholesterol and fibrinogen, a pattern similar to that observed for BMI, rather than for height, was observed. There was considerable familial aggregation, including non-trivial spouse correlations. A variety of genetic and environmental models could be fitted for each variable, depending on how the environmental effects common to each set of adult relative pairs are parameterised.

Whether one can make a definitive statement about which model gives the 'best' fit will depend on a number of factors, including the design (i.e. what sorts of relatives have been studied, what are their ages, etc.) and data collection (i.e. what measures are available to allow more sophisticated modelling of the effects of shared environment; see 8. below). Often, the only honest conclusion that can be made is to state that one cannot discriminate, even though a range of models might be fitted. That is, the conclusion of 'insufficient evidence' might be appropriate, and should not be avoided just because there might be a set of rules to choose the 'best fit' on statistical grounds alone, given the data. Statistical methods are designed to assist in making judgements and decisions, and are not meant to over-ride sensible judgements about the quality and discriminatory power of the data. Recognising that the design and data collection is inadequate is a major part of good applied statistics.

7. Confounding between the effects of shared genes and shared environment

With data from twins alone, it is well known that if the 'ACE model' is fitted – allowing for additive genetic factors (A), common environment factors (C), and individual specific factors (E) – there is a strong negative correlation between the estimate of the A effect and the estimate of the C effect. For example, if A is estimated by $2(r_{MZ} - r_{DZ})$, and C by $2r_{DZ} - r_{MZ}$, then it is a straightforward mathematical exercise to show that $\text{Corr}(A,C) = -(0.9)^{\frac{1}{2}} = -0.95$.

That is, even under the strong common environment assumption of the classic twin model, there is little statistical power to discriminate between the effects of shared genes from those of shared environments. The twins alone design can, therefore, be very limited. However the next section discusses a twins alone study which covers an important period of life, from the pre-pubertal years through to the mid-twenties, and a trait related to growth and development, bone mineral density.

8. The temporal nature of the effects of a shared environment, and the age group being studied

In the classic twin model, the effect of a shared or common environment is presumed to be a constant. However, the effects of a common environment shared by twins may increase during their upbringing, and may dissipate as they become adults and live more separate lives. At one extreme it could attenuate rapidly, while at the other it could persist.

In a recent cross-sectional study of bone mineral density (BMD) at the hip and spine in adolescent and young adult female twin pairs aged from 8 to 26 years old,[17] it was found that (a) the mean varied with age, and (b) the variance varied with age, as did the MZ and DZ pair covariances. The variances and covariances were estimated as a fourth-degree polynomial in age, using the computer software package FISHER,[18] both before and after adjusting for lean mass.

In particular, the DZ covariance/correlation exceeded one-half the MZ covariance/correlation during late adolescence. When the classic twin model was fitted, assuming the effects of common environment were independent of zygosity, there was evidence for a substantial effect of common environment in late adolescence that peaked around age 18, but disappeared by age 23. The cohabitation status of these pairs had been recorded, and whereas nearly all pairs 18 or younger were living together, all pairs 24 or older were living separately. That is, application of the classic twin model gave plausible fits that suggested that the effect of environmental effects shared by twins during adolescence dissipated rapidly once they lived apart. This model-generated hypothesis is being currently tested by analysing longitudinal BMD data from these and other pairs of a similar age range.

The twin analyses above also suggested that there were additive genetic components of variance for BMD around the ages of menarche. Furthermore, almost 50% of these genetic variance components were no longer evident once an adjustment for lean mass on mean BMD was included, particularly for BMD at the hip where the cross-sectional association with lean mass was strongest. Adjusting for lean mass reduced, but did not obliterate, the common environment effect in late adolescence. This suggested that exercise and physical activity might explain part, but not all, of the common environment effect. Furthermore, the additive genetic effect was increased in pairs older than 18, consistent with gene-environment interactions or covariation.

This application illustrates several important points. It suggests that there are quite substantial environmental effects on BMD during adolescence, when 'peak bone mass' is achieved, and that these environmental effects are shared by twins within a pair. It also suggests that the genetic variation observed from previous studies of adult twin pairs may be more sophisticated than simply additive genetic effects acting independently of other sources of variation. BMD variation may be due to a complex combination and interaction of genetic and environmental effects.

9. Are the models of the shared environment effects realistic?

The effects of shared environment, or cohabitation, may be more subtle than proposed in the classic twin model. To see this, however, it may be necessary to study non-twin siblings. For example, smoking status in adult nuclear families was analysed using the regressive logistic models proposed by Bonney.[19] These models allow for different types of dependencies within sibships, while concurrently allowing for a correlation between parents (but without necessarily attributing the sources of that correlation). It also allows for an association between parents and offspring. After allowing for substantial parental-offspring associations (equivalent to a two-fold increase in the odds of smoking for each smoking parent), there was evidence for substantial within-sibship effects.[20] The best description from amongst those tried suggested that if the next older sibling smoked the odds of smoking increased three-fold. This description was preferred to a model that proposed that the eldest sibling's smoking status was the predictor of smoking, or a model that proposed that an offspring's smoking status was equally predicted by the smoking status of each other sibling. The parsimonious description of the residual within-sibling effects (after allowing for parental effects) could not be explained by any known genetic mechanism. That is, whereas the parental transmission could be due to shared genes, or to 'cultural transmission', the additional within-sibship associations could not be attributed to genetic factors.

10. The presentation and discussion of 'findings'

Often when the fits of models are presented in the literature, focus is on point estimates, p-values, and so-called 'goodness-of-fit' χ^2 statistics. The latter are, in most applications, just weak tests of *lack* of fit, and unless the sample size is large they are virtually worthless. A major problem with a large number of published twin analyses has been their naïve and incorrect interpretation that, if a test of fit is statistically 'not significant', then this is evidence for a 'good' or 'acceptable' fit.

Although likelihood methods are used almost universally, only rarely is a variance-covariance matrix showing the correlations between estimates published or discussed in the text. Moreover, standard errors or confidence intervals are often lacking, making the reported estimates meaningless. Consequently, the Discussion sections of some papers are wasted on fanciful interpretations of perceived 'differences', when the statistical errors in estimates are so large that sampling and chance easily explain the differences.

11. Model fits depend on what is adjusted for in the mean

Variation in a trait is not defined, and therefore cannot be discussed, without reference to the mean about which that variation occurs.[21] That is, the proportion of variance attributable to genetic factors ('heritability') is not a fixed characteristic of a trait *per se*, but can differ depending on the factors taken into account in describing the mean of that trait.[10] Adjusting the mean for different factors can have a major effect on the conclusions of twin analyses.

In a cross-sectional study, BMD at the hip and lean mass were found to be associated in adult women, even after adjusting the means of both BMD and lean mass for age, having a correlation of about 0.4.[22] The cross-trait correlation within twin pairs in these two age-adjusted traits was about 0.3 in MZ pairs, and 0.1 in DZ pairs. When the means of both hip BMD and lean mass were also adjusted for height, the cross-trait correlation within an individual fell to 0.3, and the cross-trait within pair correlations for MZ and DZ pairs became almost identical, at about 0.16. These observations suggest that genetic factors may explain a considerable portion of the correlation between age-adjusted BMD and age-adjusted lean mass, and that genes that determine variation in height in adult women might be the source of that genetic covariation.

12. Reviewer and journal bias

Finally, some journals are much more inclined to certain styles of analyses, and types of conclusions. For example, the paper on BMD in young and adolescent twins,[17] which showed evidence for common environment as well as additive genetic components of variance, was rejected by the *American Journal of Human Genetics*, but accepted without fuss by the *American Journal of Epidemiology*.

DISCUSSION

Good science is not easy to do! Fitting models to prop up one's preconceived ideas is not good science. As discussed at the end of my review paper10, I suggested that perhaps a more fruitful way for biometric modelling to proceed would be to presume that the null hypothesis is that genetic factors are present, rather than not present. That is, twin researchers should try to falsify the genetic hypothesis. Our experiences in performing the analyses discussed above suggest that, in doing so, they will better appreciate and understand the strengths and weaknesses of their genetic hypotheses, and the modelling methods they are using.

Models are just models, not reality.

Having been critical of so much of twin data analyses, or more specifically their interpretations, published to date, at the invitation of the editors I would like to end on a more positive note by humbly giving some opinions about how one might, or might not, proceed.

Be skeptical.

Understand that the information is in the data, not in any pre-conceived ideas you might have, or the opinions of others, about what is 'correct'.

Don't over-interpret a fitted model, no matter how 'good' the fit may appear. Try to test the assumptions underlying the model, and be aware of those you cannot test (such as the presumed additivity of unmeasured genetic and environmental effects).

Appreciate that the modelling is of variation, and that the variance will differ according to what is considered to be the 'mean'. Try putting different measured factors into the modelling of the mean, and seeing what happens to the variance, and the estimated components of variance.

Calculate the trait correlation for different types of relatives, and if possible for different age groups, for those living together and for those living apart, and so on. Examine the correlations for any apparent trends that might aid modelling, keeping in mind however that future tests of significance may be misleading.

Understand that estimates, rather than 'p-values', are what is important. However, note that an estimate is of no value unless it is accompanied by a measure of its imprecision, such as a standard error or confidence interval.

The major aim of modelling twin and family data should not be to come up with an estimate of 'heritability'. The concept is often misunderstood[10,23], and I have deliberately avoided using the term in this paper.

Be prepared to conclude that the design and data collection are such that one cannot make definitive conclusions about the different sources of variation. It is ok to say 'I don't know'.

Understand that model fitting does not constitute 'proof'. Statement A implies statement B does not mean that statement B implies statement A. A good model fit is a necessary, but not sufficient, condition.

Be honest scientists, not apologists or protagonists for any philosophy or dogma.

Don't confuse the ability to get a computer package to print out 'results' with being a good scientist.

Be self-critical.

ACKNOWLEDGEMENTS

This work was supported by the Australian National Health and Medical Research Council (NHMRC). Many of the applications presented in this paper have arisen out of twin studies conducted using the Australian Twin Registry, and I would like to express my appreciation to the thousands of twins, and their relatives, who have contributed to those studies.

REFERENCES

1 Peto J. Genetic predisposition to cancer. In: Banbury Report 4: Cancer incidence in defined populations. Cold Spring Harbor Laboratory, 1980: pp 203–213.

2 Hopper JL, Carlin JB. Familial aggregation of a disease consequent upon correlation between relatives in a risk factor measured on a continuous scale. *Am J Epidemiol* 1992; 136: 1138–1147.

3 Khoury MJ, Beaty TH, Liang K-Y. Can familial aggregation of disease be explained by familial aggregation of environmental risk factors? *Am J Epidemiol* 1988; **127**: 674–83.

4 Fisher RA. The correlation between relatives on the supposition of Mendelian inheritance. *Trans Roy Soc Edinb* 1918; **52**: 399–433.

5 Wright S. Correlation and causation. *J Agric Res* 1921; **20**: 557–585.

6 Elston RC, Stewart J. A general model for the genetic analysis of pedigree data. *Hum Hered* 1971; **21**: 523–542.

7 Lange K, Westlake J, Spence MA. Extensions to pedigree analysis: II. Variance components by the scoring method. *Ann Hum Genet* 1976; **39**: 485–491.

8 Hopper JL, Mathews JD. Extensions to multivariate normal models for pedigree analysis. *Ann Hum Genet* 1982; **46**: 373–383.

9 Neale MC, Cardon LR. *Methodology for genetic studies of twins and families.* London: Kluwer. 1992

10 Hopper JL. Variance components for statistical genetics: applications in medical research to characteristics related to human diseases and health. *Stat Meth Med Res* 1993; **2**: 199–223.

11 Christian JC, Norton JA, Sorbel J, Williams CJ. Comparisons of analysis of variance and maximum likelihood based path analysis of twin data: partitioning genetic and environmental sources of covariance. *Genet Epidemiol* 1995; **12**: 27–35.

12 Hopper JL, Mathews JD. Extensions to multivariate normal models for pedigree analysis. II. Modelling the effect of shared environment in the analysis of variation in blood lead levels. *Am J Epidemiol* 1983; **117**: 344–355.

13 Clifford CA, Hopper JL, Fulker DW, Murray RM. A genetic and environmental analysis of a twin study of alcohol use, anxiety and depression. *Genet Epidemiol* 1984; **1**: 63–79.

14 Hopper JL, Culross PR. Covariation between family members as a function of cohabitation history. *Behav Genet* 1983; **13**: 459–471.

15 Becker, NG. *Analysis of Infectious Disease Data.* London: Chapman & Hall. 1989.

16 Harrap SB, Stebbing M, Hopper JL, Hoang NH, Giles GG. Familial patterns of covariation in cardiovascular risk factors: The Victorian Family Heart Study. (submitted).

17 Hopper JL, Green RM, Nowson CA, Young DA, Sherwin AJ, Kaymakci B, Larkins RG, Wark JD. Genetic, common environment, and individual specific components of variance for bone mineral density in 10- to 26-year old females: a twin study. *Am J Epidemiol* 1998; 147: 17–29.

18 Lange K, Boehnke M, Weeks D. *Programs for pedigree analysis.* Los Angeles: Department of Biomathematics, UCLA. 1987.

19 Bonney GE. Regressive logistic models for familial disease and other binary traits. *Biometrics* 1986; **42**: 611–625.

20 Hopper JL, Jenkins MA, Macaskill GT, Giles GG. Regressive logistic modelling of familial aggregation for smoking in a population-based sample of nuclear families. *J Epidemiol Biostats* 1997; **2**: 45–52.

21 Hopper JL. Genetic correlations and covariance. In: *Encyclopedia of Biostatistics*. London: John Wiley & Sons. 1998; **2**: 1669–1676.

22 Seeman E, Hopper JL, Young NR, Formica C, Goss P, Tsalamandris C. Do genetic factors contribute to associations between muscle strength, fat-free mass and bone density? A twin study. *Am J Physiol: Endocrin Metab* 1996; 270: E320-E327.

23 Hopper JL. Heritability. In: *Encyclopedia of Biostatistics*. London: John Wiley & Sons. 1998; **3**: 1905–1906.

14

USING SIB-PAIRS AND PARENT-CHILD TRIOS IN ASSOCIATION STUDIES

Cathryn M. Lewis

ABSTRACT

Tests for allelic association using controls from within families are being widely used as an alternative to case-control association studies. These are used to narrow the candidate regions identified through genome searches, and to identify the causality of candidate genes, and could be applied to genome-wide association studies. The transmission/disequilibrium test (TDT) is the most widely used test, and is simple to apply, requiring only the counts of the number of alleles transmitted from a heterozygous parent to an affected offspring. It is a test of linkage in the presence of association. This test has recently been extended to allow for wider conditions including multiple allele markers, quantitative traits, sib-TDT without parents and the use of additional affected and unaffected sibs. This paper provides a review of these methods and considers in detail the properties that affect the power of the TDT to detect allelic association.

LINKAGE DISEQUILIBRIUM

Linkage disequilibrium occurs where the alleles at two genetic loci are not independent, because they are tightly linked and have rarely been separated by recombinations. These loci may be causative mutations which increase the risk of developing a disease, or non-functioning marker polymorphisms. The notation used throughout this paper is that M, m denote alleles at a marker locus, and D, d denote alleles at a disease-causing locus. The loci are in linkage disequilibrium if the frequency of the haplotype DM differs from that expected by the population frequencies of alleles D and M.

POPULATION HISTORY OF LINKAGE DISEQUILIBRIUM

Mutations in DNA are rare events, and we can assume that individuals who carry a particular genetic variant are descendants of a single ancestral mutation event. Transmission of this mutation through the generations will be accompanied by transmission of flanking sections of DNA which are undisturbed by recombination events. Thus, polymorphisms in this genetic region will be in linkage disequilibrium with the disease mutation, while not being a causative factor for the disease. For example, a mutation D which occurs on the ancestral chromosome is inherited through to current generations, along with portions of the DNA from the ancestral chromosome. A marker polymorphism M, which lies close to D, will be inherited with D unless separated by recombinations, and thus M will be over-represented in affected individuals and in allelic association with the disease. The level of linkage disequilibrium between D and M depends on the ancestral frequency of the allele M prior to mutation D, and on the genetic distance between D and M. Alternatively, the polymorphism M may have arisen at a later date, and not be present on the ancestral chromosome in which the mutation D occurred. These genetic models for linkage disequilibrium between a marker, disease mutation, and their association with a disease will be discussed further in considering the factors that affect the power to detect association.

Linkage: close localisation of two genetic loci (markers or genes) implying that the alleles at these loci are not inherited independently within families.

Linkage disequilibrium: two loci are in linkage disequilibrium if they are linked, and the alleles at these loci are not independent, i.e. presence of allele 1 at locus 1 increases (or decreases) the probability of finding allele 1 at locus 2.

Association: a genetic locus is associated with a disease, if the presence of a particular allele at the locus increases the probability that an individual is affected with the disease.

TESTING FOR ASSOCIATION: LIMITATIONS OF CASE-CONTROL STUDIES

Case-control studies may be used to test for an association between a disease and a polymorphism by testing the equality of the proportion of cases and controls who carry the polymorphism. Although this is a well-defined epidemiological study design, there are pitfalls in its applications to genetics which have yet to be overcome, and many significant results have never been replicated in further studies, or fail to hold when a further set of controls is analyzed.

The underlying problem lies in matching cases and controls on their genetic background. Information on geographic location of the most recent ancestors and ethnicity can be useful, but is not always available or sufficient. Biases can arise as follows: consider a population which is comprised of two underlying equally sized populations A and B which do not inter-marry and a rare disease with a higher prevalence in population A than population B. In a case-control study, the cases will have an excess of members from population A, and the controls be approximately mixed from both populations. Therefore any genetic association study comparing allele frequencies in cases and controls is, in effect, comparing allele frequencies in population A and population B, which may be independent of the disease for which the samples were originally collected. While the example given here is overly simplistic, it illustrates how population stratification may bias study results.

TESTING FOR ASSOCIATION: WITHIN-FAMILY CONTROLS

These problems with population stratification can be overcome by using related controls who will be better matched genetically. Various methods have been developed, including the transmission/disequilibrium test,[1] haplotype relative risk[2], and affected family-based controls.[3] The most widely used of these is the transmission/disequilibrium test (TDT). This test in its original classical form as described by Spielman, McGinnis and Ewens[1] uses a family unit of a single affected offspring and two parents, with genotype data available from each individual. Figure 1 shows a TDT-trio with an affected offspring, and the parental alleles for a marker with alleles M and m. Each heterozygous parent contributes to the test, with the 'case' being the allele which is transmitted to the affected offspring, and the 'control' the untransmitted allele. These data can be summarised into a contingency table of the transmitted and non-transmitted alleles, and tested for departures from random transmission using McNemar's test to give a chi-square statistic with one degree of freedom.

The TDT measures distortion of the inheritance of alleles from parents to an affected offspring. In Figure 14.1, if allele M is in linkage disequilibrium with the disease mutation, it will be transmitted in excess to affected offspring, whilst if alleles M and m are not in linkage disequilibrium with a mutation for the disease, transmission will occur at random. The test is robust to population stratification and makes no assumption of Hardy- Weinberg equilibrium. The TDT is a test of both linkage and linkage disequilibrium, so the null hypothesis may take either form. However, linkage will only be detected if allelic association is present, so that the distortion of transmission acts on the same allele in all families. Similarly, allelic association will only be detected in the presence of genetic linkage and not where it is due to population stratification.

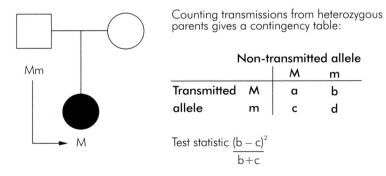

Counting transmissions from heterozygous parents gives a contingency table:

		Non-transmitted allele	
		M	m
Transmitted	M	a	b
allele	m	c	d

Test statistic $\dfrac{(b-c)^2}{b+c}$

Figure 14.1 – TDT description and statistic

EXTENSIONS OF THE TDT

One attraction of the TDT is its simplicity but several extensions have been developed in order to allow application of the TDT to wider practical situations.

1. Multiple alleles

The original derivation of the TDT was for a biallelic marker. This permits testing of the presence or absence of a putative disease polymorphism, or single-nucleotide polymorphism, but is not appropriate where a multiallelic marker is used. Data for the TDT from a marker with n alleles M_1, M_2, M_3, ... M_n are the number of transmissions of alleles M_i and M_j from M_iM_j heterozygous parents, giving an $n \times n$ contingency table. This problem now is to test for significant transmission distortion, and identify the subset of alleles in which that distortion occurs. Where one allele has previously been identified as potentially associated with the disease, this allele may be compared to all other alleles pooled, in the framework of the original TDT. Otherwise, methods using all marker alleles must be considered. The simplest method is to form separate contingency tables for each allele versus all other alleles pooled, and perform the original TDT within each table, using a Bonferroni correction to allow for multiple testing. Any of the statistical methods for contingency table analysis may be applied to these data, and suggested methods have included log linear models, testing for difference in row and column totals and Monte Carlo simulation.[4,5,6] The TDT with multiple alleles allows identification of several marker alleles which may be in linkage disequilibrium, either through the occurrence of several mutations on different ancestral chromosomes, or through recombinations.

2. Quantitative traits

Two recent papers have extended the TDT to be used for quantitative traits, where the offspring is selected on the basis of a continuous trait measurement,[7 8] and the following discussion of methods for quantitative traits are based on Allison.[8] In sampling families for a quantitative trait, offspring may be randomly selected, giving a population-based distribution of trait measurements. Alternatively, individuals with an extreme trait value, lying within predetermined tails for the trait distribution may be selected. Extreme sampling gives increased power to detect linkage or association, as has previously been found for sib pair linkage sharing statistics[9]. TDT-type statistics for quantitative traits use the trait values for individuals with heterozygous parents. The method compares trait values in those receiving allele M (or D) with those receiving allele m (or d). One of the simplest approaches is to compare the number of transmissions of allele M to offspring with trait values in the lower and the upper tails of the distribution. This method differs from the discrete TDT in using both tails, not only the 'affected' individuals from one tail of the distribution, but it loses power by not considering the full quantitative trait values of offspring. The most powerful statistic suggested by Allison[8] is to regress the trait value on the genotype of the affected individuals, while including dummy variables for the parental mating types. The method uses the full quantitative trait values, allows either population sampling or sampling of individuals from only the upper and lower tails, and the regression can be extended to allow for environmental variables. This should be a powerful and flexible method for detecting linkage disequilibrium in continuous traits.

3. Multiple affected siblings

The TDT detects association in the presence of linkage, and is a valuable tool in narrowing genetic regions which have been identified through genome-wide linkage searches. However, affected sib pairs are often used in linkage searches, and in using the TDT with only a single affected offspring, the information on the second sibling is lost. The TDT may be applied to multiple affected offspring, and the statistic becomes the number of transmissions of allele M measured independently to all siblings, regardless of the numbers of affected siblings within a family. However, the TDT applied to multiple affected siblings is only valid as a test of linkage, and not a test of association, since transmissions to affected offspring are not independent in the presence of linkage.[10] The TDT is valid as a test of association when a single affected sibling is selected from each family. The presence of a second affected sibling can increase the power to detect linkage disequilibrium since these families are now chosen to be more heavily genetically predisposed compared to random affected offspring selected from the population (see Power section below for examples).

4. Family structure

The TDT uses family units where one offspring and both parents are genotyped. This can be restrictive, particularly for late-onset traits where parents are not available for sampling. Although it is possible to infer genotypes for missing parents, this may lead to biases induced by the non-independence of the parent-to-offspring transmitted allele and the missing parental genotypes.[11] A more appropriate solution for dichotomous traits is the use of unaffected siblings; the genotypes of the affected sibling is compared to that of the unaffected sibling without attempting to reconstruct parental genotypes. Several methods have been proposed. Spielman and Ewens[12] develop the sib-TDT, with a statistic which compares the number of M alleles in affected offspring compared to unaffected offspring. This allows families with differing numbers of affected and unaffected siblings to be included. Curtis[13] allows for the availability of several unaffected siblings, but uses only the sibling whose genotype is most different to that of the affected offspring (e.g. MM and mm). Boehnke and Langefeld[14] use a discordant sib pair (one affected, one unaffected), and provide power calculations for the number of discordant alleles between these siblings across a range of genetic models.

Although these methods differ in details, the basic statistic is the number of M alleles identified in affected siblings and in unaffected siblings. The power for these statistics is lower than for the TDT, and there are two sources for the loss of power. Firstly, unaffected siblings may still be genetically predisposed but unaffected. Secondly, the TDT is fully informative provided at least one parent is heterozygous, whilst a sibling-based version of the TDT is only informative if the (untyped) parent is heterozygous AND has transmitted both alleles to offspring. Distributions for these sib-control statistics are not easy to determine analytically and may be obtained through Monte Carlo simulations or permutation methods.[12,14] The application of these sibling control tests is less straightforward than the original TDT, but the ability to test for association using only siblings will be a valuable tool relevant to many genetic studies with family resources of this type.

STUDY DESIGN ISSUES

As we have shown above, the TDT provides a useful method to detect linkage disequilibrium within samples which are available (or easily collected) in family studies. However, the robustness against possible population stratification may have a severe cost in the numbers of samples required.

The factors which will affect the power of the TDT to identify susceptibility genes are:

1. Role of polymorphism

Detecting linkage disequilibrium with the disease-causing mutation requires a smaller sample size than using a marker allele which is in linkage disequilibrium with the mutation. This may be parametrised by the increased risk conferred by the genotypes DD and Dd compared to the baseline risk for genotype dd. For a marker with alleles M, m in linkage disequilibrium with the mutation, the risks associated with the genotypes MM, Mm and mm will depend on both the underlying disease mutation risks and the linkage disequilibrium between the marker and the mutation. Unless the high risk marker allele is in complete linkage disequilibrium with the mutation, the genotypic relative risks from the marker alleles will be less discriminating than those for the mutation.

2. Frequency of polymorphism

The TDT detects distortion from random inheritance of parental alleles, i.e. departures from 50% inheritance of allele M from heterozygous (Mm) parents to affected offspring. The sample size required to detect this distortion depends on the underlying frequency of the polymorphism, in addition to the increase in the transmission probability. Figure 14.2 shows the expected number of informative parent-offspring pairs from TDT-trios required to obtain 80% power to obtain a significant departure from random inheritance for an increased transmission of 55% – 65% of allele M (with a type I error of 0.05). The heterozygosity of the polymorphism is important in determining the proportion of the total sample that will be informative with at least one heterozygous parents. The 'full information' line represents parent-offspring pairs in which all parents are heterozygous, which is a theoretical maximum information level. In practice, the number of heterozygous parents is maximised with a polymorphism of 50% frequency, and polymorphisms of only 5–10% frequency will require much larger sample sizes for the same distortion of transmissions. The increase in power with an increasing allele frequency assumes the same effect size (transmission distortion). However, a disease-causing mutation is likely to have low frequency, but a substantially larger transmission distortion effect. In this situation, the gain in information from using the mutation outweighs the loss of information from the majority of parents who will be homozygous for the normal allele. Note that these calculations conservatively assumes that parents are sampled randomly from the population. Parental genotypes should be sampled conditional on their affected offspring, but this requires the (unknown) underlying genetic model.

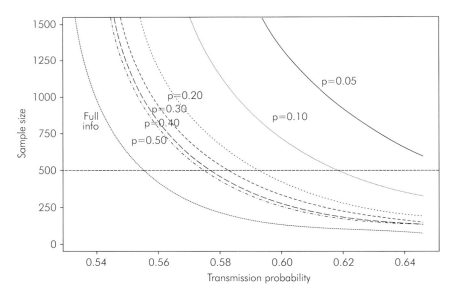

Figure 14.2 – Sample sizes of parents required to detect distortion of transmission to offspring, assuming full information (from heterozygous parents), or using allele frequencies of 0.05 – 0.5 to calculate the probabilties of heterozygous parents.

3. Affected family members

One method to increase the power of the TDT to detect linkage disequilibrium is to preferentially select families in which other affected family members exist, in addition to the affected offspring used in the analysis. Further affected siblings may be included in the analysis for linkage (but not for linkage disequilibrium), as described above. Affected parents can decrease the sample size required, whilst not increasing the number of samples to be genotyped.[15] The precise value of these further affected individuals depends on the underlying genetic model at the locus in linkage disequilibrium with the disease, but across a wide range of genetic models the presence of further affected individuals increases the probability that at least one parent is heterozygous for the mutation under test, and will contribute to the TDT statistic.

POWER CALCULATIONS

Power calculations for any method to localise genes for complex traits or phenotypes rely on an assumption of the underlying genetic effect of any locus to be detected. Little information on the genetic model can be obtained from phenotypic analysis within families, as any genetic effect will contribute only a proportion of the variability of the measurable phenotype. Sample sizes for testing for a prespecified effect for a dichotomous trait were given in Figure 14.2 above. A more thorough approach would take into account the underlying mode of inheritance of the mutation. This was provided by Risch and Merikangas[16] for multiplicative genetic models, and extended by Camp[17] to additive, recessive and dominant models. These models assume that a disease mutation is available for testing, and has alleles D and d. Models are parametrised by the frequency p of allele D, and the risk ratio γ for high risk genotypes. So, for a dominant model, the probability that an individual of genotype DD (or Dd) is affected is increased by a factor γ from the probability that an individual with genotype dd is affected. Models in table 1 have risk factors of $\gamma = 4, 2$, and 1.5 (i.e. give successively less genetic models), and gene frequencies of p=0.01, 0.1, 0.5 and 0.8. All the genetic models have only a moderate genetic component, giving a risk ratio to siblings of $1.0 - 1.5$.

Table 14.1 (reprinted from 17) gives the numbers of parent-offspring trios which should be sampled to obtain 80% power to detect a significant effect of the mutation under test. This table illustrates two features of determining power for testing linkage disequilibrium. For many of the polymorphism frequency and genetic model combinations shown, 500 TDT trios are sufficient to localise susceptibility genes. However, for certain combinations, much larger numbers are required. For example with a polymorphism frequency of 10% and a genotypic risk ratio of 4, less than 200 families give sufficient power provided the high risk allele acts in a multiplicative, additive or dominant fashion, but over 5000 trios are

Table 14.1 – Sample sizes required for TDT trios with a single affected offspring, and two affected offspring, under multiplicative, additive, recessive and dominant genetic models. Table taken from Camp[17] (copyright permission obtained from the University of Chicago press).

Frequency	Singletons				Affected sib pairs			
	Multiplicative	Additive	Recessive	Dominant	Multiplicative	Additive	Recessive	Dominant
γ = 4.0								
0.01	523	549	4.3×10^6	562	59	67	7.0×10^5	70
0.10	86	123	5056	153	20	36	866	49
0.50	103	222	205	712	71	143	104	516
0.80	291	663	337	9873	293	518	312	7,810
γ = 2.0								
0.01	4,154	4,154	3.8×10^7	4,317	1,018	1,018	1.2×10^7	1,083
0.10	533	533	43,331	766	161	161	13,762	268
0.50	340	340	949	1,861	188	188	431	1,133
0.80	750	750	976	22,728	553	553	661	15,149
γ = 1.5								
0.01	16,008	15,550	1.5×10^8	16,487	5,366	5,121	6.1×10^7	5,579
0.10	1,908	1,485	1.7×10^5	2,554	704	490	67,104	1,012
0.50	949	464	3,078	4,599	492	237	1,423	2,575
0.80	1,833	816	2,553	51,914	1,156	583	1,509	31,206

required for a recessive model. The genetic model is an unknown factor which (unlike the polymorphism frequency) cannot be estimated through a pilot study and so some mutations will remain undetectable using the TDT. A stringent significance level of 5×10^{-8} is used in table 1, which is appropriate for testing thousands of polymorphisms in a genome-wide association study. Sample sizes would be reduced considerably for testing a few candidate polymorphisms, although the variation in sample sizes across genetic models would still hold.

The sample sizes required for families with two affected offspring are substantially lower than those for a singleton. In almost all models, the reduction in sample size justifies the genotyping of the additional sibling, and the reduction can be substantial (for example 533 to 161 families for $\gamma = 2$, $p = 0.1$ and a multiplicative or additive model). However, the TDT applied to both siblings in an affected sib pair is now a test for linkage, not linkage disequilibrium. The importance of this distinction depends on the context in which the TDT is performed. Where regions identified through a genomic search are being densely genotyped, linkage disequilibrium is required to narrow the region, and therefore the linkage statistic obtained from applying the TDT to sib pairs is less useful. However, in genome-wide association studies any indication of linkage or association will be valuable, and both sibs should be used in the analysis.

Table 1 assumes that the disease-causing mutation is available for testing. For testing a marker polymorphism in linkage disequilibrium with the disease, the sample sizes will generally increase, and the change will depend on the marker allele frequency and the extent of disease-marker disequilibrium.[18] A more subtle assumption in interpreting these power calculations concerns the genetic model. The model assumed applies only to the mutation which is tested, not to the gene in which the mutation occurs. This implies that linkage disequilibrium is much easier to detect where a single ancestral mutation is responsible for all the excess risk attributed to a particular gene. Where several mutations exist in a gene, the genotype relative risks attributable to a single mutation will be smaller than that due to the gene, and sample sizes required will increase accordingly. Many of the published studies into power of the TDT and its derivative tests are for qualitative traits, however the underlying factors that affect power (polymorphism frequency and genetic model) apply equally to quantitative traits. Simulation studies of power of the TDT to detect association for quantitative traits have been performed by Allison[8] and Rabinowitz.[7]

CONCLUSION

Genetic studies have advanced from Mendelian diseases to complex traits where risk is attributable to both genetic and environmental factors, and the genetic contribution will arise from many genes each having small effect. The TDT has proved a valuable tool in identifying the causality of candidate gene polymorphisms and narrowing regions identified in linkage analysis. It should also have high power to identify mutations in genome-wide association studies and this will become apparent as the necessary molecular technology develops. No single uniformly powerful method will exist for identifying these genes and their interactions with other genetic or environmental factors. High-risk families, TDT-trios, and case-control series, covering both isolated and outbred population groups, will all be required for this exciting but daunting task of dissecting the genetic etiology of common diseases.

REFERENCES

1 Spielman RS, McGinnis RE, Ewens WH: Transmission test for linkage disequilibrium: the insulin gene region and insulin-dependent diabetes mellitus (IDDM). Am J Hum Genet 1993; 52:506- 526

2 Falk CT, Rubinstein P: Haplotype relative risks: an easy reliable way to construct a proper control sample for risk calculations. Ann Hum Genet 1987; 51:227–233

3 Thompson G: Mapping disease genes: family-based association studies. Am J Hum Genet 1995; 57:487–498

4 Sham PC, Curtis D: An extended transmission/disequilibrium test (TDT) for multi-allele marker loci. Ann Hum Genet 1995; 59:97–105

5 Bickeböller H, Clerget-Darpoux F: Statistical properties of the allelic and genotypic transmission/disequilibrium test for multi-allelic markers. Genet Epidemiol 1995; 12:865–870

6 Morris A, Curnow RN, Whittaker JC: Randomisation tests of disease-marker association. Ann Hum Genet 1997; 61:49–60

7 Rabinowitz D: A transmission disequilibrium test for quantitative trait loci. Hum Hered 1997 47:342–350

8 Allison DB: Transmission-disequilibrium tests for quantitative traits. Am J Hum Genet 1997; 60:676–690

9 Risch N, Zhang H: Extreme discordant sib pairs for mapping quantitative trait loci in humans. Science 1995; 268: 1584–1589

10 Spielman RS, Ewens WJ: The TDT and other family-based tests for linkage disequilibrium and association. Am J Hum Genet 1996; 59:983–989

11 Curtis D, Sham PC: A note on the application of the transmission disequilibrium test when a parent is missing. Am J Hum Genet 1995; 56:811–812

12 Spielman RS, Ewens WJ: A sibship test for linkage in the presence of association: the sib transmission/disequilibrium test. Am J Hum Genet 1998; 62:450–458

13 Curtis D: Using siblings as controls in case-control association studies. Ann Hum Genet 1997; 61: 319–333

14 Boehnke M, Langefeld CD: Genetic association mapping based on discordant sib pars: the discordant-alleles test. Am J Hum Genet 1998; 62:950–961

15 Whittaker JC, Lewis CM: The effect of family structure on linkage tests using allelic association. Am J Hum Genet 1998; 889–897

16 Risch N, Merikangas K: The future of genetic studies of complex human diseases. Science 1996; 273:1516–1517

17 Camp NJ: Genomewide transmission/disequilibrium testing–consideration of the genotypic relative risks at disease loci. Am J Hum Genet 1997; 61:1424–1430

18 Müller-Myhsok B, Abel L: Genetic analysis of complex diseases. Science 1997; 275:1327–1330

15

THE CONCEPT OF GENOME-WIDE POWER AND A CONSIDERATION OF ITS POTENTIAL USE IN MAPPING POLYGENIC TRAITS: THE EXAMPLE OF SIB-PAIRS

David B. Allison
Myles S. Faith

ABSTRACT

There is growing interest in mapping genes for polygenic complex traits. However, obtaining sufficient sample sizes to have adequate power can be quite challenging even when selectively sampling extreme relative pairs. The task of obtaining sufficient sample size can shift from challenging to Herculean when one conducts a random screen of the genome and appropriately controls the genome-wide type I error rate to the nominal α level (1). Herein, we describe the concept of genome-wide power (or its complement, genome-wide Type II error), defined as the probability of obtaining significant linkage to one or more of k influential loci. We show that as long as k, the number of influential loci, is > 1 then genome-wide power is always > the power to detect linkage for all or each of the k loci. Finally, it is proposed that assuring sufficient genome-wide power rather than locus-specific power may, in some cases, be considered an acceptable standard for the design of linkage studies of polygenic traits and disorders.

Currently there is increasing interest in using linkage studies to identify genes for complex polygenic traits and disorders. When dealing with complex traits, one typically does not know with certainty the genetic mode of inheritance. Therefore, it is generally considered advisable to use a statistical approach that does not depend on a particular genetic model. Perhaps the most common sampling unit used with such approaches is the sibling pair. The "model free" approach has the advantage of making no assumptions about the mode of inheritance. Unfortunately, this approach and the nature of polygenic complex traits in general make it very difficult to achieve adequate statistical power.

In this paper, we draw heavily upon two prior papers. One is a paper by Lander and Kruglyak[1] that discusses issues of significance testing in the context of genome scans. They present the concept of genome-wide α level, i.e., the probability of making one or more type 1 errors across the whole genome under a global null hypothesis. By global null hypothesis, we mean that there is no marker linked to the phenotype in the whole genome. The second paper, by Risch and Zhang,[2] illustrates the sample sizes necessary to achieve a given degree of statistical power when using extremely discordant sibling pairs (EDSP'S) for mapping quantitative trait loci (QTL's). They present formulas for adjusting sample sizes as a function of other variables (significance level, desired power, recombination, marker informativeness). As discussed below, formulae and tables used in Risch and Zhang's paper were adapted herein.

Consider the following example. An investigator wishes to identify genes for human body fatness, where human body fatness will be operationalized by the body mass index (BMI; kg/m^2). Assume further that the investigator is looking for loci that explain ten percent of the variance in BMI and have a population frequency of 0.30. Finally, assume that, independent of the effects of the loci in question, there is a residual correlation among siblings of 0.40 and a bivariate normal residual distribution. Under these circumstances, one of the most powerful designs available is to use extremely discordant sibling pairs.[3,4] For the current paper, we will define EDSP'S as sibling pairs in which one sibling has a BMI greater than or equal to the 90th percentile and the other sibling has a BMI less than or equal to the 10th percentile. Using formulae and tables from Risch and Zhang,[2] it can be shown that the number of EDSP's needed to have 80% power (at the one-tailed .05 level) under these conditions and assuming a recessive mode of inheritance is 116. However, as Risch and Zhang point out, this assumes perfect marker heterozygosity and a recombination fraction of zero. Assuming more realistically that one will conduct a multi-point analysis[5–9] with markers situated at 20 centiMorgan intervals in the genome and having an average polymorphism information criteria (PIC) of 0.70 then the required number of EDSP's is 116x2.26=262. Furthermore, given that only about 0.8% percent of sibling pairs will meet this definition of extremity[10], then the expected number of pairs that need to be screened equals 262/.008 = 32,750. Finally, it should pointed out that these calculations assume that parents are available and fully typed which, in reality, is often not the case.

Although screening over 32,000 pairs of siblings and phenotyping and genotyping 262 pairs would be a large job, for a phenotype such as BMI it is probably realistic. However, as random screening of the genome involves a great multiplicity of testing, the inflation of the α rate (Type I error probability) is quite great. Therefore, Lander and Kruglyak[1] recently proposed guidelines for significance testing with complete screens of the genome. Specifically, they provided formulae for determining the point wise α level necessary to achieve a genome wide type I error probability of no greater than α. When the genome-wide α level is set at 0.05, then the point-wise α level equals 2.2x10^{-5}. Using formulae from Risch and Zhang[2] the number of EDSP's needed to achieve eighty percent power under the same assumptions as above but with the point-wise α level set to 2.2x10^{-5} rather than 0.05, is approximately 1,048 and therefore the expected number to be screened is approximately 131,000. As can be seen, the use of such a significance level in genome screening results in an enormous increase in the necessary sample size (approximately 4-fold) that, even with selective sampling strategies, makes obtaining a sufficient sample size a truly Herculean task.

In the remainder of this paper, the simple but important concept of genome-wide power (and its complement, genome-wide type II error) is introduced as a natural counterpart to the concept of genome-wide type I error. Methods of estimating genome-wide power and an example of its use are provided. It is proposed that the use of genome-wide power in sample size calculations can markedly reduce the number of pairs necessary to have adequate power and, under certain circumstances, may be considered an acceptable alternative in the design of linkage studies for polygenic traits and disorders.

THE CONCEPT OF GENOME-WIDE POWER

Let there be k loci influencing a polygenic trait or disorder . Each of the i loci has a specific heritability (h^2_i) and therefore, its own power level denoted $1-\beta_i$.

Genome-wide then, there are three kinds of power to consider. The first is the power to find *all* loci significant. This power is simply $\Pi_i(1-\beta_i)$, where $\Pi_i(X_i)$ denotes the running product of the X_i.★ The second type of power is the power to find a *specific* locus significant. For the ith locus, this power is simply $1-\beta_i$. Finally, there is the power to find *any* locus significant. We refer to this as the genome-wide power. The genome-wide power is given by the quantity $1- \Pi_i(\beta_i)$. Because each of the β_i are bound between 0 and 1.0, it can be seen that, for any β_i, $\Pi_i(1-\beta_i) < 1-\beta_i < 1-\Pi_i(\beta_i)$, as long as k is greater than 1. Therefore, calculating the required sample size on the basis of genome-wide power will always result in a smaller required sample size than basing the required sample size on the loci specific powers which seems to be the traditional approach. Thus, basing sample size on genome-wide power may be a useful alternative in certain contexts.

EXAMPLE REVISITED

Returning to the example above, assume that in fact there were three loci influencing BMI, each of which had a locus specific heritability of 0.10. Then, setting the genome-wide power to be equal to 0.80 means $0.80 = 1-\Pi_i(\beta_i)$. Solving for β_i indicates that β_i is equal to $0.2^{1/3} = .58$. In other words, the genome-wide power in this scenario equals 0.80 when the locus specific power is set at $1-0.2^{1/3} = .42$. Using this desired locus specific power in the formulae provided by Risch and Zhang[3] and maintaining α at 2.2×10^{-5}, then it is seen that the desired number of

---

★The derivation assumes the loci are independent.

EDSP's is 650. This represents a reduction of 38% in necessary sample size. It should be noted that the power to detect all significant loci (the first type of power) is only .07 (that is, 0.42^3).

DETERMINING THE NUMBER OF INFLUENTIAL LOCI AND BI

The example provided was simplified by knowing the value of k and knowing that the locus specific heritability was equal for all values of k. In reality, these two things are very unlikely to be known.

To estimate the value of k, several alternatives might be used. First, methods are available to conduct this estimation among inbred strains of animals.[11-13] In linkage studies involving laboratory animals, these methods would be applicable. One might (*with great tentativeness*) extrapolate such a value to humans. Second, although there may be limitations to their use, methods for estimating the number of genes influencing a complex trait also exist for humans.[14] This method combined with the application of segregation analyses may provide reasonable guidelines. Third, as with many of the parameter estimates needed for power analysis, one might ultimately provide a reasonable guess. Finally, because most of the parameters fed into any power analysis are simply informed guesses it seems reasonable to calculate power over a plausible range of k and present all of the results.

CONCLUSION

In conclusion, for any given study design, the number of observations needed to achieve a genome-wide power of $1-\beta$ would be reduced relative to the number required to achieve a power of $1-\beta$ for each locus of interest. If each locus explained a similar amount of variance and had a similar mode of inheritance then the required sample size would be reduced by a factor of

$1 - [(Z_{(1-\beta)}1/k_j - Z_\alpha)/(Z_{(1-\beta)} - Z_\alpha)]^2$, where Z_X is the standard normal deviate that cuts off the upper 100X percent of the normal distribution. This formula is derived by adaptation of equation 3 from Risch and Zhang.[2] Table 15.1 indicates the reduction in the necessary sample size for values of k from 2 to 10 with α set to 2.2×10^{-5} from the guidelines of Lander and Kruglyak[1] and β set to 0.2 as is fairly common practice.

In this paper, the concept of genome-wide power was introduced. Genome-wide power was defined as the probability of obtaining *any* significant linkage in a study investigating a polygenic trait influenced by k separate loci. Though, the concept is made explicit here, it has been implied previously ([15], p. 849). Moreover, it underlies the observation of Suarez et al.[16] that the expected sample size required to detect linkage to one of several oligogenes is less than the expected sample size

Table 15.1 – Percent reduction in necessary sample size as a function of k (the number of influential loci) when sample size calculations are based on genome-wide (i.e., the third type of power) rather than locus-specific power (i.e., the second type of power). Assumes that desired power is .80 and that α is set at 2.2×10^{-5}.

k	$1 - .2^{1/k}$	$Z_{(1 - .2^{1/k})}$	Percent Reduction
2	0.55	0.13	27%
3	0.42	0.20	38%
4	0.33	0.44	45%
5	0.28	0.58	49%
6	0.24	0.71	53%
7	0.21	0.81	56%
8	0.18	0.92	59%
9	0.16	0.99	60%
10	0.15	1.04	62%

required to replicate linkage to the particular locus initially detected.

It is now recognized that steps must be taken to insure that the genome-wide α rate stays fixed at a reasonable level. However, insuring this in a random genome screen may increase the necessary sample size to a prohibitive degree. Therefore, it is advocated that investigators simultaneously consider the concept of genome-wide power and consider this as one possible standard on which to judge the acceptability of the power of proposed linkage studies of complex polygenic traits.

REFERENCES

1   Lander E, Kruglyak L. Genetic dissection of complex traits: guidelines for interpreting and reporting linkage results. *Nature Genet.* 1995, **11**, 241–247.

2   Risch N, Zhang H. Mapping quantitative trait loci with extreme discordant sib pairs: sampling considerations. *Am J Hum Genet* 1996, **58**, 836–843.

3   Eaves L, Meyer J. Locating human quantitative trait loci: guidelines for the selection of sibling pairs for genotyping. *Beh Genet* 1994, **24**, 443–455.

4   Risch N, Zhang H. Extreme discordant sib pairs for mapping quantitative trait loci in humans. *Science* 1995, **268**, 1584–1589.

5   Fulker DW, Cherny SS, Cardon LR (1995) Multipoint interval mapping of quantitative trait loci, using sib pairs. *Am J Hum Genet*, **56**, 1224–1233.

6   Goldgar DE. Multipoint analysis of human quantitative genetic variation. *Am J Hum Genet.* 1990, **47**, 957–967.

7   Kruglyak L, Lander ES. Complete multipoint sib-pair analysis of qualitative and quantitative traits. *Am J Hum Genet* 1995, **57**, 439–454.

8   Olson JM. Multipoint linkage analysis using sib pairs: an interval mapping approach to dichotomous outcomes. *Am J Hum Genet.* 1995, **56**, 788–798.

9   Schork NJ. Extended multipoint identity-by-descent analysis of human quantitative traits: Efficiency, power, and modeling considerations. *Am J Hum Genet*, 1993, **53**, 1306–1319.

10   Allison DB. The use of discordant sibling pairs for finding genetic loci linked to obesity: practical considerations. *Int J Obes.* 1996, **20**, 553–560.

11   Cockerham CC. Modifications in estimating the number of genes for a quantitative character. *Genetics* 1986, **114**, 659–664.

12   Ollivier L, Janss LLG. A note on the estimation of the effective number of additive and dominant loci contributing to quantitative variation. *Genetics 1993,135*, 907–909.

13   Zeng Z. Correcting the bias of Wright's estimates of the number of genes affecting a quantitative character: A further improved method. *Genetics* 1992,**131**, 987–1001.

14   Risch N. Linkage strategies for genetically complex traits. *Am J Hum Genet* 1990, **46**, 222–228.

15   Smalley SL, Woodward JA, Palmer CGS. A general statistical model for detecting complex-trait loci by using affected relative pairs in a genome search. *Am J Hum Genet,* 1996, **58**, 844–860.

16   Suarez BK, Hampe CL, Van Eerdewegh P. Problems of replicating linkage claims in psychiatry. In: Gershon ES, Cloninger CR, eds. *Genetic Approaches to Mental Disorders*, 1993, Washington DC: American Psychiatric Press, 1993, pp. 23–46

16

THE USE OF TWINS IN QUANTITATIVE TRAIT LOCUS MAPPING

Nicholas J. Schork
Xiping Xu

ABSTRACT

The exploitation of DNA markers in mapping loci that influence variation in quantitative traits can be pursued with a variety of statistical analysis methods. However, there are a few fundamental concepts that are common to all mapping analysis strategies. These fundamental concepts bear on the need to correlate some measure of genotypic similarity at a particular locus with a measure of phenotypic similarity among a group of related individuals. If such a correlation exists, then, it is arguable that variation at the locus in question is likely to influence variation in the phenotype. The use of twins has a long and illustrious history in the investigation of genetic and non-genetic factors contributing to phenotypic correlations. In this paper we show how one can exploit genotype-phenotype correlations among and between twins to yield a powerful and flexible approach to mapping and characterizing genes that influence quantitative variation in humans. We contrast the proposed twin-based approach with standard mapping approaches based on sibling pairs, and discuss issues surrounding the implementation details of the proposed approach.

The use of twins in quantitative genetic analysis has been long-standing and highly productive (see, for example, the discussion in section 9.11 of the classic textbook by Cavalli-Sforza and Bodmer[1]). The contrast between monozygotic (MZ) twins, who share all their genes, and dizygotic (DZ) twins, who, like other non-MZ siblings, share half of their genes (on average), with respect to similarities in phenotypic features offers an orderly setting for investigating the effect of gross genetic and environmental factors on phenotypes that exhibit quantitative or continuous variation in the population at large. In addition, by collecting and accommodating information on the environment in which the twins used for an analysis have been reared, greater insight into the role of the environment in producing variation in the phenotype of interest can be obtained.

One genetics research area where twins have not been used extensively is in mapping *specific* or *individual* genes that influence quantitative traits such as blood pressure and cholesterol level (although there are some notable exceptions[2]). This is somewhat odd for two reasons. First, many standard approaches to gene mapping involve the assessment of allele sharing patterns among sibling pairs.[3–5]

Since DZ twins can be considered age-matched sibling pairs, their use in standard sibling pair-based gene mapping analyses seems ideal. Second, since one can exploit the contrast between MZ and DZ twins to better characterize the effect of genetic and environmental factors on variation in a trait, one could conceivably exploit this contrast to better characterize the role of *a particular locus* with respect to variation in a phenotype.

In this paper we describe an analysis approach to gene mapping for quantitative traits that makes use of twin data. The proposed approach is derived within a simple linear mixed model framework. We contrast this proposed model with a standard sibpair-based model for quantitative trait locus mapping. We also consider issues relating to the inclusion in the model of factors that might influence the trait of interest beyond the locus being mapped, as well as possible hypothesis testing strategies. We ultimately show that the proposed models have great potential to aid in the genetic and environmental dissection of quantitative traits.

THE HASEMAN-ELSTON MODEL FOR SIBPAIR ANALYSIS

Basic Construction

Before describing our twin model for gene mapping, we outline the basic concepts behind a traditional approach to gene mapping that uses sibling pairs. We do this to provide a contrast with our proposed approach. The traditional approach was first outlined by Haseman and Elston some 25+ years ago.[6] This procedure involves regressing the squared sibpair difference in trait values,

$$d = (Y_1 - Y_2)^2$$

(a measure of phenotypic similarity, where Y_1 and Y_2 are trait values of the 1st and 2nd sibling) on the fraction of alleles shared identical-by-descent by the sibpair at the locus in question, $\hat{\pi}$, (a measure of genotypic similarity) where a decrease in d with increasing $\hat{\pi}$ is interpreted as evidence for linkage. If i indicates the *ith* sibling pair out of a total of N sibpairs, then a simple linear regression model relating d to $\hat{\pi}$, can be fashioned:

$$d_i = b_0 + b_1\hat{\pi}_i + e_i \tag{1}$$

where b_0 is an intercept term, b_1 is a regression coefficient, and e is an error term with mean 0 and variance σ^2. The fraction of alleles shared identical-by-descent at a locus, $\hat{\pi}$, can be estimated from marker genotype data collected at loci surrounding the locus of interest.[7] The intercept, slope, and error parameters in equation (1) can be estimated via least squares. A test of linkage involves the null hypothesis $H_0 : b_1 = 0$ against the alternative $H_1 : b_1 < 0$. A standard one-sided t-test can be used for testing purposes, as can non-parametric rank correlation tests.[6]

This procedure could easily be used with twins, whereby monozygotic twins would automatically assume a value of 1.0 for $\hat{\pi}$, whereas dizygotics would have $\hat{\pi}$, estimated from marker data.[2]

Problems with the Haseman-Elston Model

Despite its patent elegance and simplicity, the Haseman-Elston (HE) procedure as described above has a number of shortcomings. The following is a brief list of these shortcomings:

- Does not account for environmental exposures/correlations explicitly
- Can not accommodate covariates (e.g., gender) easily or intuitively
- Can not accommodate multiple locus effects easily
- Awkward distributional properties (i.e., d ~ distributed as folded-normal if the underlying trait is normally distributed)
- Has poor power properties in unselected samples
- Does not make maximal use of the sibling pair trait values

Of these problems the last two may be the most important. Wright recently showed that consideration of the correlation between $\hat{\pi}$ and the mere difference in sibling trait values, $d = y_1 - y_2$ ignores information inherent in the complete bivariate distribution of y_1 and y_2 and therefore results in a less powerful test of linkage than one that makes use of the complete joint distribution of y_1 and y_2.[8] In fact, Wright showed that the expected lod score for a linkage test using just the difference could be increased by the factor:

$$F = 1 + ELOD_S / ELOD_D \approx 1 + (\frac{1 - \rho_0}{1 + \rho_0})(\frac{1 - \rho_2}{1 + \rho_2})$$

where $ELOD_S$ and $ELOD_D$ are expected lod scores from regressions of the sibling trait value *sum* and *difference* on $\hat{\pi}$, respectively, and where ρ_0 and ρ_2 are correlations between sibling trait values for siblings that share 0 and 2 alleles IBD, respectively. This factor could be quite large for small correlations, thus making the HE procedure less powerful for loci with small effect. (Note that Gaines and Elston[9] derived the same equation for a probability measure of zygosity given phenotype data from twins 30 years prior to Wright's work!).

Fortunately, one can indeed make full use of the sibling trait values through the use of standard variance component or linear mixed effects models.[10] In addition, such models can accommodate covariates, environmental factors, and multiple locus effects. Finally, in recent work, Goldgar,[11] Goldgar and Oniki,[12] Schork,[13] and Fulker et al[14]. have all shown, through simulation studies, that variance components models are more powerful than the standard Haseman-Elston procedure in the *general* analysis of sibling pairs. We therefore propose variance component models[15] in the analysis of twin data for QTL mapping purposes.

VARIANCE COMPONENT MODELS AND TWINS

The Basic Model

Let y_1 and y_2 denote quantitative trait values collected from a twin pair. Assume, for the time being, that the twin pair trait value vector, $Y = [y_1,y_2]$ can be modeled with an appropriate bivariate distribution (e.g., bivariate normal) with mean vector, μ, and variance-covariance matrix, Ω which can be partitioned in the following way:

$$\Omega = \sum_{l=1}^{L} \Pi_l \sigma_l^2 + 2K\sigma_a^2 + H\sigma_c^2 + I\sigma_e^2 \qquad (2)$$

where σ_l^2, σ_a^2, σ_c^2 and σ_e^2 are estimable variance components terms characterizing locus specific, residual (i.e., non-marked or non-directly measured locus) additive genetic, shared household or environment, and random environmental effects, respectively. The coefficient terms preceding the variance terms are 2×2 coefficient matrices relating the variance components to the twin pair trait values. Thus, Π_l is an identity-by-descent allele sharing matrix computed via marker genotypes obtained at loci flanking the locus in question,[7, 16] K is the kinship coefficient matrix (15), C is a matrix characterizing shared households or environments,[17] and I is the identity matrix. Table 16.1 lists the elements of the coefficient matrices used in the construction of equation.[2]

Assume further that the trait mean, μ, can be modeled as $\mu = f(X|B)$, where X is a vector of covariates (e.g., gender, age, etc.) and B is an estimable parameter vector. The variance component terms and the parameter vector B can be estimated via

Table 16.1 –The 2×2 Coefficient Matrices Used in the Construction of the Twin Pair Covariance Matrix.

Coefficient Matrix		Monozygotic Twins	Dizygotic Twins
Π_l		1 1 1 1	1 π_l π_l 1
$2K$		1 1 1 1	1 ½ ½ 1
C	No Environment Sharing:	1 0 0 1	1 0 0 1
	Environment Sharing:	1 1 1 1	1 1 1 1
I		1 0 0 1	1 0 0 1

maximum likelihood. If bivariate normality of Y is assumed, the relevant log-likelihood equation is:

$$L(B, \sigma_1^2, ..., \sigma_L^2, \sigma_a^2, \sigma_c^2, \sigma_e^2 \mid Y, X) = -\tfrac{1}{2} \log|\Omega| - \tfrac{1}{2} (Y - BX)' \, \Omega^{-1} \, (Y - BX) \qquad (3)$$

If more than one twin pair is collected, the log-likelihood equation becomes the sum of the individual log-likelihoods for each twin pair.

Two things should be noted about this proposed model. First, L locus effects can be modeled simultaneously, although in practice one may be interested in a single locus effect. Second, in some cases the variance component parameters might be virtually unidentifiable (in a statistical sense). For example, if all the twins have been reared apart, then the C and I matrices will be identical for all twin pairs. Similarly, if uninformative markers are used, all $\hat{\pi}$ values will equal ½ and thus the Π and $2K$ matrices will be identical for all twin pairs. In these cases a reduction in the number of variance components included in the model can be made, although not without a loss in ability to make comprehensive interpretations of the analysis results.

Testing Procedures

To test for a locus effect while accounting for the effects of other factors, one can perform likelihood ratio tests. Assume for the time being that $L=1$ (i.e., there is only one locus being tested). Then one can compute a log-likelihood using equation 3 assuming no locus effect (the null hypothesis), $H_0 : \sigma_1^2 = 0$, and a log-likelihood assuming a locus effect (the alternative hypothesis), $H_1 : \sigma_1^2 > 0$. Minus twice the ratio of these log-likelihoods gives a statistic that should be distributed asymptotically as a half: half mixture of a χ^2 statistic with 1 degree of freedom and a point mass at 0 (due to the constraint the $\sigma_l^2 \geqslant 0$ see [18]).

An alternative testing procedure would be to conduct profile tests, as outlined by Goldgar.[11] This testing strategy works in two stages in the following way. Consider a single locus model with variance terms $\sigma_l^2, \sigma_a^2, \sigma_c^2$ and σ_e^2. Interest is in the effect of the locus, l, on the trait. The testing stages are:

Stage 1: Estimate via maximum likelihood a mean parameter, μ, and variance terms σ_a^2, σ_c^2 and σ_e^2.

Stage 2: Fix the parameters estimated in stage 1 to their values and maximize, via maximum likelihood estimation, a parameter, α, which quantifies *how much of the additive genetic variance the locus explains*. (i.e., $\sigma_l^2 = \alpha \sigma_a^2$) The relevant covariance decomposition becomes:

$$\Omega = \Pi_l(\alpha \sigma_a^2) + 2K(1 - \alpha)\sigma_a^2 + C\sigma_c^2 + I\sigma_e^2$$

This procedure could be used with simpler models initially. For example, one could simply ask how much of the overall variance (σ_p^2) the locus explains and thereby ignore (or simply not accommodate in the model) other sources of variation. Thus, the covariance decomposition would be:

$$\Omega = \Pi_l(\alpha\sigma_p^2) + I(1-\alpha)\sigma_p^2$$

This procedure would allow the use of previously obtained estimates of certain factors and avoid computational problems inherent in the estimation of multiple parameters. It does, however, ignore gene x environment and other forms of interaction.

Modeling and Power Considerations

There are many things worth examining in the variance component approach to twin-based gene mapping and general genetic analysis. We focus on a few issues, recognizing that more work should be done. We first investigated properties of the profile likelihood ratio test for assessing linkage described in the previous section. As discussed, this testing strategy requires fitting a model that does not assume a specific locus effect (stage 1), and then fitting a model that assumes a specific locus plus all the factors assumed in stage 1 (stage 2). The model fit in stage 2, however, fixes the parameters estimated in stage 1 to their estimated values and merely assess how much variation explained by the factors assumed in stage 1 might be due to the specific (linked) locus effect. The relevant likelihood ratio statistic investigating $H_0: \alpha = 0$ (vs. the one-sided alternative $H_1: \alpha > 0$) may have a null distribution that is influenced by:

- Sources of variation not accounted for in the model
- Major locus effect not explicitly accommodated in the model

We investigated the effect of these phenomena via simulation. Three stage 1 models were investigated. These included models accommodating the following effects (in terms of variance components):

- Model 1: σ_e^2
- Model 2: σ_a^2, σ_e^2
- Model 3: $\sigma_a^2, \sigma_c^2, \sigma_e^2$

500 simulations were run with 200 twin pairs (25% monozygosity) assuming different residual additive, shared environment, random environment, and unlinked major locus effect contributions to the trait, but no linked locus effect. The purpose of these simulations is to determine if the probability of erroneously concluding that a specific locus influences a trait is increased (or decreased) when models that do not accommodate all relevant factors contributing to the trait are fit to the data. Tables 16.2 and 16.3 offer the results of these simulations as well as a comparison with the HE procedure (Table 16.2). In addition, an assessment of the accuracy of the parameter estimates is offered (Table 16.3). From Table 16.2 it can

Table 16.2 – Comparison of Theoretical and Empirical Type 1 Error Rates for the Variance Component Model Profile Test and the Haseman-Elston Regression Test, Based on 500 Simulations.

Effect Sizes			VC (σ_e^2)		VC (σ_c^2, σ_e^2)		VC ($\sigma_a^2, \sigma_c^2, \sigma_e^2$)		HE	
σ_a^2	σ_c^2	σ_e^2	0.050	0.010	0.050	0.010	0.050	0.010	0.050	0.010
0	0	1	0.038	0.008	0.000	0.000	0.026	0.002	0.044	0.008
			2.371	3.829	0.290	0.701	1.705	3.190	−1.561	−2.317
0	1	1	1.000	1.000	0.000	0.000	0.016	0.000	0.048	0.006
			55.787	64.886	0.884	1.579	1.422	3.451	−1.636	−2.306
1	0	1	0.976	0.864	0.004	0.000	0.042	0.002	0.106	0.024
			26.513	31.500	1.589	2.378	2.480	3.978	−2.012	−2.627
1	1	1	1.000	1.000	0.004	0.000	0.046	0.014	0.102	0.022
			64.265	74.616	1.412	2.230	2.550	5.958	−1.987	−2.619

Key: Effect sizes = variation explained by the listed factor, VC = variance component model (with assumed stage 1 variance components); HE = Haseman-Elston Regression. Upper entries are emprical type 1 error rates associated with a chi-squared statistic for variance component models (critical values of 2.71 and 5.41 for 0.05 and 0.01 type 1 error rates, respectively) and one-sided t-statistics for the Haseman-Elston regression model (critical values of −1.65 and −2.33), Lower entries are the order-statistic derived critical values obtained from the simulations. 200 twin pairs were assumed with fully informative flanking markers 1 cM away from the assumed trait-influencing locus. Probability of monozygosity was set at 0.25.

be seen that when additional factors beyond the locus being tested influence the trait, but these factors are not accounted for in the model, the type I error rates are inflated. For example, if the total variance of the trait was equal to 3.0, with each variance component contributing equally (i.e., $\sigma_a^2 = 1, \sigma_c^2 = 1, \sigma_e^2 = 1$ the last row of Table 16.2) then a model that only allows for a locus effect and a residual effect (i.e. the first two columns of Table 16.2 under VC(σ_e^2)) erroneously leads to the inference of a locus effect almost all the time (note that the HE procedure will have increased type I error rate as well), although a model with all components accounted for (i.e., VC(σ_a^2, σ_c^2, σ_e^2)) will produce the expected error rates. In addition, from Table 16.3 it can be seen that a model with shared household effects will erroneously attribute variation to shared households if a residual genetic effect exists but is not accounted for in the model (last line of Table 16.3, under VC(σ_c^2, σ_e^2)). Table 16.4 describes type I error rates, average parameter estimates, and comparison with the HE procedure, when one uses a model that accounts for all sources of variation and uses a standard likelihood ratio test. It can be seen from Table 16.4 that a complete model controls type I error rates well. That is, the observed type 1 error rates and estimated order statistics from the test statistic distribution obtained over the simulations all match what one would expect. Although, as can be seen from the 'VC parameter estimates' columns, parameter estimates for shared household and residual additive genetic variance are confounded (i.e., they are far from their expected values, especially as indicated in the scenarios associated with the last two rows of Table 16.4). This is most likely the result of the identifiability problems discussed in the section on 'Modeling and

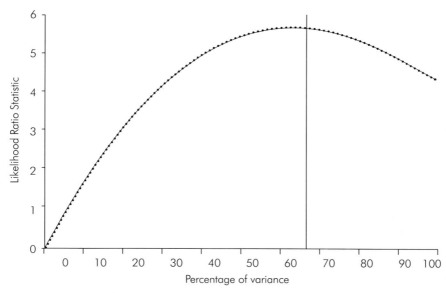

Figure 16.1. – Profile likelihood curve for a locus effect obtained from a single simulation involving 200 twin pairs and a locus effect that explains 66% of the trait variation.

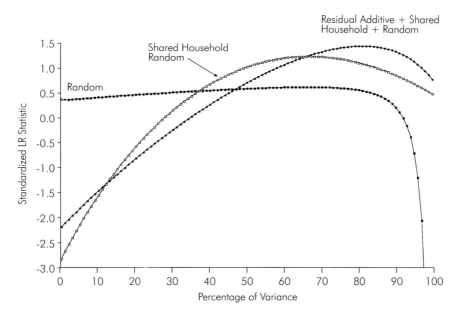

Figure 16.2 – Profile likelihood curves for a locus effect when different models are fit to a single simulated data set (see text for explanation of the setting used to simulate the data).

Table 16.3 – Average Parameter Estimates Obtained for the Variance Component Models Used in the 500 Simulations Investigated for the Construction of Table 16.2.

Effect Sizes			VC (σ_r^2)	VC (σ_c^2, σ_e^2)		VC (σ_a^2, σ_c^2, σ_e^2)		
σ_a^2	σ_c^2	σ_e^2	σ_e^2	σ_c^2	σ_e^2	σ_a^2	σ_c^2	σ_e^2
0	0	1	0.994	0.026	1.007	0.248	0.130	0.987
			0.0031	0.0017	0.0044	0.0153	0.0091	0.0130
0	1	1	1.990	0.995	0.995	0.277	0.993	0.993
			0.0073	0.0071	0.0043	0.0159	0.0185	0.0148
1	0	1	1.993	0.545	1.443	1.003	0.183	1.022
			0.0064	0.0060	0.0060	0.0290	0.0120	0.0146
1	1	1	3.004	1.543	1.443	1.057	0.990	1.004
			0.0110	0.0110	0.0063	0.0300	0.0211	0.0148

Key: (see Table 16.2). Upper entries are the average parameter estimate obtained over the 500 simulations; the lower entry is the standard error of the parameter estimates. All parameter estimates given in the table were computed under the null model of no locus effect. Parameter estimates under the alternative were at times markedly different, leading to the increased false-positive rates described in Table 16.2.

Table 16.4 – Comparison of Theoretical and Empirical Type 1 Error Rates for a Four-Parameter Variance Component Model Test and the Haseman-Elston Regression Test Based on 500 Simulations.

Effect Sizes			VC (σ_a^2, σ_c^2, σ_e^2)		HE		VC Parameter Estimates		
σ_a^2	σ_c^2	σ_e^2	**0.050**	**0.010**	**0.050**	**0.010**	σ_a^2	σ_c^2	σ_e^2
0	0	1	0.046	0.004	0.062	0.018	0.139	0.135	0.997
			2.465	3.865	−1.800	−2.478	0.0097	0.0090	0.0089
0	1	1	0.050	0.010	0.042	0.010	0.214	1.285	0.665
			2.622	5.297	−1.534	−2.330	0.0250	0.0148	0.0100
1	0	1	0.066	0.010	0.114	0.034	0.247	0.951	1.004
			3.145	4.693	−2.103	−2.701	0.0153	0.0230	0.0128
1	1	1	0.066	0.016	0.102	0.028	0.323	2.118	0.690
			3.416	6.066	−2.023	−2.555	0.0214	0.0327	0.0127

Key: Effect sizes = variation explained by the listed factor, VC = variance component model (with assumed stage 1 variance components); HE = Haseman-Elston Regression. Upper entries are empirical type 1 error rates associated with a chi-squared statistic for variance component models (critical values of 2.71 and 5.41 for 0.05 and 0.01 type 1 error rates, respectively) and one-sided t-statistics for the Haseman-Elston regression model (critical values of −1.65 and −2.33), Lower entries are the order-statistic derived critical values obtained from the simulations. 200 twin pairs were assumed with fully informative flanking markers 1 cM away from the assumed trait-influencing locus. Probability of monozygosity was set at 0.1. For "VC Parameter Estimates" the upper entry is the average parameter estimate and the lower entry is the standard error.

Power Considerations'. Thus, one should always err in the direction of including more in the model and eliminate insignificant factors as revealed only through testing and finding them to be not significant statistically.

The power to detect locus effects under various settings was also investigated by simulation. Figures 16.1 and 16.2 offer graphs of profile likelihoods computed under different models, assuming that a locus explains 66% of the trait variation and the rest is equally attributable to a shared twin household effect and a random effect. 200 twin pairs were assumed with a fully informative marker. 10% of the twins were assumed to be MZ and 25% were assumed to share the same household. Figure 16.1 shows the profile of the locus effect with a model that only accounts for a locus effect, the household effect, and a random effect. Figure 16.2 shows profiles when other effects are included or ignored. Figure 16.1 shows that the profile is maximized near an assumed locus effect that accounts for 66% of the trait variation, which is to be expected given that this was the assumption used to generate the data. However, if models that do not account for factors that influence the trait are fit to the data (i.e., Figure 16.2), then the profile for the locus effect is not maximized at the appropriate place.

The likelihood ratio statistic investigating $H_0: \alpha = 0$ (vs. the one-sided alternative $H_0: \alpha = 0$) was investigated via simulation for other settings as well. Many factors could influence power:

- The linked locus effect size (locus-specific heritability)
- Distance of markers from the locus
- Informativity of the markers

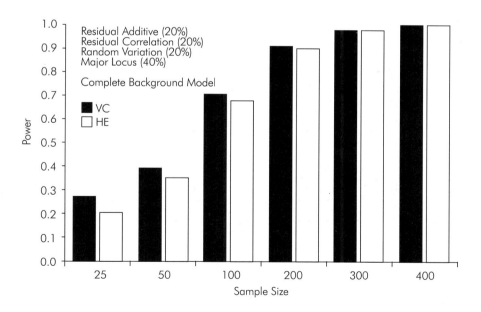

Figure 16.3. – Power of the variance component approach as a function of sample size compared to the HE approach when a model assuming locus effect, residual genetic, shared household, and random effects are fit to simulated data.

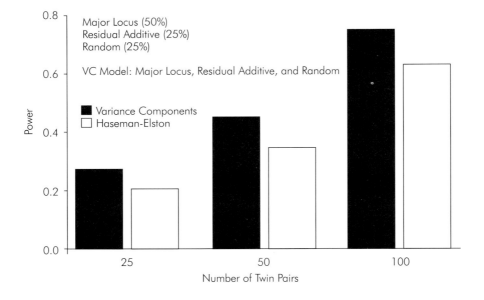

Figure 16.4. – Power of the variance component approach as a function of sample size compared to the HE approach when a model assuming locus effect, residual genetic, and random effects are fit to simulated data.

- Effect of other sources of variation
- Number of twin pairs
- Fraction of twin pairs that are monozygotic
- Fraction of twin pairs sharing environments

Since it would be difficult to assess the impact of all these factors on power, we choose to examine a few scenarios. We again used the stage 1/stage 2 profile testing strategy discussed previously. We first examined the effect of sample size (i.e., number of twin pairs) in a single hypothetical situation. A fully informative marker was assumed in the analyses. A model assuming residual additive, shared household (25% of the twins were assumed not to share households), and random effects, in addition to a locus effect, were fit to the simulated data The effect of sample size on power is described in Figure 16.3 for this scenario. We also investigated the effect of sample size on power in a hypothetical situation in which no common environment effect was assumed. The results are described in Figure 4. It can be seen from Figures 16.3 and 16.4 that the variance component model can be more powerful than the standard HE model in certain settings.

DISCUSSION

This paper has tried to make clear how easily one can construct models for gene mapping and general genetic analysis that make use of twin pairs. Although not without some problems relating to just what one should account for in a relevant mapping model, the proposed approach should provide a more insightful analysis strategy than traditional gene mapping strategies based on sibling pairs or large pedigrees.[5] Further extensions of the proposed twin approach to gene mapping and genetic analysis could involve the use of whole genome parameters,[19] the use of information provided by the extended families of twin pairs, selected samples,[20] or the accommodation of qualitative traits.[21] Ultimately, although the use of twins was once reserved for the mere estimation of the contribution of gross environmental and genetic factors to phenotypic variation (an exercise now thought to be somewhat outdated in an era of molecular biology), their use in gene mapping and gene effect characterization studies should become more pronounced as genotyping and sequencing capabilities improve.

ACKNOWLEDGEMENTS

Work by NJS is supported, in part, by United States National Institutes of Health grants HL94–011, HL54998–01, and RR03655–11 and the financial generosity of GEMINI Research Ltd of Cambridge, England.

REFERENCES

1 Cavalli-Sforza LL, Bodmer WF. The Genetics of Human Populations. San Francisco: W.H. Freeman and Company, 1971.

2 Cardon LR, Smith SD, Fulker DW, Kimberling WJ, Pennington BF, DeFries JC. Quantitative trait locus for reading disability on chromosome 6. Science 1994;266: 276–279.

3 Lander ES, Schork NJ. Genetic dissection of complex traits. Science 1994;265: 2037–2048.

4 Schork N, Chakravarti A. A nonmathematical overview of modern gene mapping techniques applied to human diseases. In: Mockrin S, ed. Molecular Genetics and Gene Therapy of Cardiovascular Disease. New York: Marcel Dekker, Inc., 1996: 79–109.

5 Schork NJ, Xu X. Sibpairs versus pedigrees: What are the advantages? Diabetes Reviews 1997;5: 116–122.

6 Haseman JK, Elston RC. The investigation of linkage between a quantitative trait and a marker locus. Behavior Genetics 1972;2: 3–19.

7 Fulker DW, Cardon LR. A sib-pair approach to interval mapping of quantitative trait loci. American Journal of Human Genetics 1994;54: 1092–1103.

8 Wright F. The phenotypic difference discards sib-pair QTL linkage information. American Journal of Human Genetics 1997;60: 740–742.

9 Gaines RE, Elston RC. On the probability that a twin pair is monozygotic. American Journal of Human Genetics 1969;21: 457–465.

10 Searle SR, Casella G, McCulloch CE. Variance Components. New York: John Wiley, 1992.

11 Goldgar DE. Multipoint analysis of human quantitative variation. American Journal of Human Genetics 1990;47: 957–967.

12 Goldgar DE, Oniki RS. Comparison of a multipoint identity-by-descent method with parametric multipoint linkage analysis for mapping quantitative traits. American Journal of Human Genetics 1992;50: 598–606.

13 Schork NJ. Extended multipoint identity-by-descent analysis of human quantative traits: Efficiency, power, and modeling considerations. Am. J. Hum. Genet. 1993;53: 1306–1319.

14 Fulker DW, Cherny SS. An improved multipoint sibpair analysis of quantitative traits. Behavior Genetics 1996;26: 527–532.

15 Lange K, Westlake J, Spence MA. Extensions to pedigree analysis. III. Variance components by the scoring method. Annals of Human Genetics 1976;39: 485–491.

16 Kruglyak L, Lander ES. Complete multipoint sib pair analysis of qualitative and quantitative traits. American Journal of Human Genetics 1995;57: 439–454.

17 Astemborski JA, Beaty TH, Cohen BH. Variance components analysis of forced expiration in families. American Journal of Medical Genetics 1985;21: 741–753.

18 Self SG, Liang KY. Asymptotic properties of maximum likelihood estimators and likelihood ratio tests under non-standard conditions. Journal of the American Statistical Society 1987;83: 605–610.

19 Schork NJ, Theil B, St. Jean P. Linkage analysis, kinship, and the short-term evolution of chromosomes. Journal of Experimental Zoology 1998 281: 133–149

20 Beaty TH, Liang KY. Robust inference for variance components models in families ascertained through probands. I. Conditioning on proband's phenotype. Genetic Epidemiology 1987;4: 203–210.

21 McCulloch CE. Maximum likelihood variance components estimation for binary data. Journal of the American Statistical Association 1994;89: 330–335.

17

MULTIVARIATE QTL ANALYSIS USING STRUCTURAL EQUATION MODELLING: A LOOK AT POWER UNDER SIMPLE CONDITIONS

Dorret I. Boomsma
Conor V. Dolan

ABSTRACT

In linkage analysis of quantitative, complex, traits the power to detect loci that explain a small to medium proportion of the genetic variance is problematic. In this paper we address the question how genetic analysis of multivariate data can be employed to increase the power to detect a quantitative trait locus using identity-by-descent mapping in sibling pairs, or dizygotic twins. These analyses are carried out with structural equation modeling, using the Mx computer program. Power calculations show that structural equation modeling is superior to the Haseman and Elston regression method. Furthermore, the power to detect a QTL can be substantially increased by considering multiple indicators of the phenotypic trait of interest. In the models used the gain in power beyond three or four indicators was, however, minimal. Detection of a dominant gene effect was shown to be unrealistic because of the large numbers needed.

Structural equation modeling or genetic covariance structure modeling (GCSM), provides a general and flexible approach to analyze data gathered in genetically informative samples[1,2]. In applying GCSM to such data, genotypic and environmental effects are modeled as the contribution of latent (unmeasured) variables to the (possibly multivariate) phenotypic individual differences. These latent factors represent the effects of many unidentified influences. In the case of a genetic factor, these effects are due to a possibly large, but unknown, number of genes (polygenes). The contributions of the latent variables are estimated as regression coefficients in the linear regression of the observed variables on the latent variables. Given an appropriate design, providing sufficient information to identify these regression coefficients, actual estimates may be obtained using a number of well disseminated computer programs, such as LISREL[3], or Mx[4] (Neale, 1997). These programs allow estimation of parameters by means of a number of estimators including normal theory maximum likelihood (ML) and weighted least squares (WLS). The latter can be applied to analyze correlations among discrete variables (e.g. tetrachoric or polychoric correlations) and nonnormal variables. A very useful estimator in the Mx program is the normal theory raw data likelihood estimator. This estimator enables one to handle missing data and to model selected samples.

Identification of quantitative genetic models is achieved, for example, by inclusion of monozygotic (MZ) and dizygotic (DZ) twins into the study. MZ twins are genetically identical while DZ twins (and siblings) share on average 50% of their segregating genes. If MZ twins are found to resemble each other more closely than DZ twins, this suggests that genetic influences are contributing to the phenotypic individual differences in the trait under consideration. One advantage of GCSM is that this approach can be generalized readily to multivariate and longitudinal data. Just as twin data can be used to decompose the variance for a single trait into a genetic and a non-genetic part, multivariate twin data can be used to decompose the covariance between traits, or between repeated measures of the same trait, into a part due to genetic covariance and a part due to environmental covariance between variables[1,5].

DETECTION OF QTLS

The flexibility of GCSM is also evident in the relative ease with which observed genotypic or environmental information can be incorporated into the analysis. An important recent development involves the incorporation of genetic information derived from marker data, which makes it possible to detect quantitative trait loci[6-8]. A quantitative trait locus (QTL) represents a stretch of a chromosome, which includes a segregating gene that contributes to individual differences in the phenotype of interest. The segregating gene has a relatively large contribution to the phenotypic variance compared to the contributions of each polygene making up a genetic latent variable. However, compared to the total effects of the polygenetic and environmental effects, the effect of the QTL may be quite small. For instance, the QTL may account for a mere 5%, or 10% of the phenotypic variance. In GCSM, the QTL is treated in the same way as a polygenetic or an environmental factor, i.e., as a latent variable. The relationship between the QTL and the phenotypic individual differences is also modeled using linear regression. The correlation between QTL factors of siblings is obtained from measured genotypic (marker) data.

The simultaneous analysis of DNA marker data and phenotypic information from sib-pairs, or dizygotic twins, to test for the presence of a QTL was developed by Haseman and Elston[9]. In addition to the measured phenotype in the sib-pairs, the Haseman and Elston method requires data relating to the siblings' genotypes at specific loci in the vicinity of the QTL. Such loci serve as markers, i.e., genetic polymorphisms with known and detectable alleles. Using the marker data, it is possible to establish the expected proportion of alleles at a given marker locus that the sibs share identical by descent (IBD, see below). The Haseman and Elston method[9] involves regressing the squared phenotypic difference score of the sibs on this proportion.

The detection of a QTL has been viewed as problematic, because of its expected relatively small effect size, and the requirement of extensive (and expensive) marker data. However, several developments have made QTL analysis feasible: the availability of marker sets consisting of many highly informative markers distributed throughout the genome (and the increasingly cheap methods of marker typing); the development of multi-point mapping methods to obtain optimal estimates of IBD status throughout the genome[10-12]; the development of selective sampling strategies to identify the most informative sib-pairs[13-18]; and, finally, the replacement of the Haseman and Elston regression method with genetic covariance structure modeling[6-8].

The use of GCSM, instead of the Haseman and Elston regression method, allows one to model the effects of a single QTL on the bivariate distribution of the sib-pairs, and to simultaneously analyze multiple indicators of a given phenotype[6]. Analyzing the bivariate distribution instead of the squared phenotypic sib-pair difference score has been shown to be more powerful[7]. As the use of multiple indicators is known to increase power in factor analysis to detect a latent factor[8], it is likely that the multiple indicators will also increase the power to detect the presence of a QTL.

MODELS

In this chapter we investigate how the use of multivariate data, compared to univariate data, increases the statistical power to detect a QTL. Multivariate data can be collected by measuring the same variable at different time points or by measuring different (correlated) variables at the same time point. The present power calculations supplement those presented in Boomsma and Dolan[20]. Boomsma and Dolan [20] considered 3 and 4 indicator models and two linear combinations of the indicators. Their calculations are limited to a codominant QTL. Here we also consider a 4 indicator model. However, we consider a dominant QTL in addition to a codominant QTL, and we investigate the effects on the power of introducing additional indicators to the model. The specific design that we focus on in this chapter is one in which the same trait is measured repeatedly across time. We assume that the time-interval between measurement occasions is short and that observed phenotypic individual differences are due to the same genes (QTL and background genetic effects) at each time-points, and that no new genetic influences are expressed across time. Measurement error (or time-specific environmental influences) thus is the only source of discontinuity across time.

Before introducing the models employed in the power calculations, we explain briefly the meaning of the term 'identity by descent' (IBD), as this is a central concept in QTL analysis. The two parents of a sib-pair are characterized by two alleles at each marker locus (say, A_i, A_j and A_k, A_l). Each member of a sib-pair inherits a single allele from his mother (A_i or A_j) and a single allele from his father (A_k or A_l). The sib-pairs may both have inherited the same allele from their father and the same allele from their mother (e.g. A_iA_k and A_iA_k). In this case, the sib-pair is characterized by IBD status 2 at the marker locus. Alternatively, the sibs may share the same allele from the mother (Ai), but each sibling inherited a different allele from the father (A_k and A_l; resulting in genotypes A_iA_k and A_iA_l in the offspring). They are then characterized by IBD status 1. Finally, they may have inherited a different allele from the father and a different allele from the mother. In this case they are IBD 0 at the marker locus (A_iA_k and A_jA_l). The reader is referred to Table 17.2 in Haseman and Elston[9] for an exhaustive list of possibilities. Note that:

1. IBD status is a characteristic of a sib-pair, not of an individual sibling;
2. IBD status at a given marker may be hard, if not impossible, to establish if, for example, alleles of the parents are identical (see Haseman and Elston[9] Table 17.2);
3. a parent and child have by definition IBD status 1 and MZ twins have IBD status 2 across all loci;
4. IBD status tells you nothing about the actual genotype of the sib-pairs.

If a marker is situated at a large distance from the QTL, the IBD status at the marker locus will be uninformative of the IBD status at the QTL due to recombination. However, if the marker is close to the QTL, the IBD status at the marker locus can serve as a proxy for the IBD status at the QTL. The IBD status at the marker locus can then be used to determine the degree of genetic relatedness at the QTL, just as the degree of genetic relatedness between additive polygenetic values of sib-pairs is expressed by the correlation of 0.5. It is this information that is exploited in both the Haseman and Elston regression method and in structural equation modeling methods to identify the regression coefficients in the regression of the phenotype(s) on QTL.

In practice, the marker data of the sibs and, if available, from their parents, are used to estimate the proportion of alleles shared IBD by the sibs (e.g. Kruglyak and Lander, 1995). These proportions corresponding to IBD=0, IBD=1 and IBD=2, are 0, 0.5 and 1 respectively. The probability that a sibpair shares a specific proportion of alleles IBD (either p[0], p[½], or p[1]) is calculated for each sibpair conditional on their marker data. The unconditional values of these probabilities, i.e. the expected values in the population, equal p[0]=.25, p[½]=.5, and p[1]=.25.

IBD marker probabilities provide information about the contribution of the QTL to the phenotypic resemblance of the sib-pair. If the QTL is codominant, the correlation between the QTL effects of sibpair i is equal to the estimate of the mean proportion of alleles shared IBD in sibpair i, π_i, and can be given by:[21]

$$\pi_i = p[½]_i{}^\star.5 + p[1]_i.$$

The effects of a dominant QTL are modeled in two parts: an additive part and a dominant part. The additive part is represented by the so-called breeding value and the dominant part, by the dominance deviation. The correlation between the sibs in breeding value still equals π_i, but the correlation of the dominance deviations equals $p[1]_i$. In summary, the correlation between the latent variables of the sibpair i are:

latent variable	correlation
polygenic additive latent factor	0.5
unshared environmental factor	0
additive QTL part (breeding values)	π_i
QTL dominance deviation	$p[1]_i$

In practice, both π_i and $p[1]_i$ may vary between 0 and 1 (although values of π_i do constrain values of $p[1]_i$, and vice versa). As explained below, we introduce simplifying assumptions, that result in π_i and $p[1]_i$ assuming a limited number of values. This greatly facilitates the power calculations. To indicate the expected (population) value of $p[1]_i$ and π_i, we drop the subscript i. These values are $\pi = .5$ and $p[1] = .25$.

Haseman and Elston Model

The original Haseman and Elston[9] sib-pair approach to linkage analysis with quantitative traits estimates the regression of the squared difference between trait values of siblings on the proportion of alleles shared IBD at a marker locus:

$$Y_i = \alpha + \beta\pi_i.$$

Let P(i,j) denote the zero mean phenotypic scores of sib j (j=1,2) in sibship i (i=1,N), then Y_i equals $[P(i,1) - P(i,2)]^2$, and π_i is the proportion of alleles shared IBD by the sibs in sibship j at the marker locus. If the regression is negative and significant, this is evidence for linkage. If there is no recombination between the marker and the QTL locus, β is a direct estimate of -2Vq, where Vq is the variance attributable to the QTL. The expectation for the squared difference score of two siblings, E[Y], may be written as:

$$E[Y] = Var(P(i,1)) + Var(P(i,2)) - 2cov(P(i,1), P(i,2))$$
$$= 2(Ve + Va + Vq) - 2(0.5Va + \pi\, Vq)$$
$$= 2Ve + Va + 2Vq - 2\,\pi\, Vq, \tag{1}$$

where Ve denotes variance due to environmental effects not shared by family members, Va denotes variance due to background genetic effects, Vq denotes the QTL variance. It may be seen from this expression that, when working with squared difference scores, Ve and Va are not separately identified.

If we consider the possibility that the effect of the QTL on the phenotype consists of an additive (codominant) genetic component and a non-additive (recessive or dominance) part, the expectation for Y can be written as:

$$E[Y] = 2Ve + Va + 2Vq + 2Vd - 2\pi\, Vq - 2p[1]Vd, \qquad (2)$$

where Ve again denotes environmental variance, Va, the variance due to background genetic effects, and Vq and Vd now represent additive and non-additive genetic variance attributable to the QTL. Equation (1) is usually fitted by ordinary least squares, and the significance of the parameter β is established by means of the t-test.

GENETIC COVARIANCE STRUCTURE MODELING INCLUDING A QTL

A structural equation modeling approach to QTL analysis with univariate sib-pair data involves the model:

$$P(i,j) = \lambda_a A(i,j) + \lambda_e E(i,j) + \lambda_q Q(i,j) + \lambda_d D(i,j) \qquad (3)$$

where $P(i,j)$, is a function of the sibs additive QTL value (Q), non-additive QTL value (D), the scores on the latent genetic background (A) and on the environmental factor (E). A path diagram for this model is given in Figure 17.1. In this model, we assume that all variables have zero mean. We also assume that the latent variables (A, E, Q, D) are standardized, so that the phenotypic variances and covariances only depend on the regression coefficients $(\lambda_a, \lambda_e, \lambda_q, \lambda_d)$. Finally, we assume that the latent variables are uncorrelated. As shown in equations 4 and 5, these parameters express the influence of the latent variables on the phenotype.

$$Var[P(i,1)] = Var[P(i,2)] = \lambda_a^2 + \lambda_e^2 + \lambda_q^2 + \lambda_d^2 \qquad (4)$$

$$Cov[P(i,1),P(i,2)] = 0.5\,\lambda_a^2 + \pi\,\lambda_q^2 + p[1]\,\lambda_d^2 \qquad (5)$$

This model is usually fitted using a program for covariance structure modeling, such as LISREL[3] or Mx[4]. If the phenotypes are approximately normally distributed, maximum likelihood estimation can be used and the significance of the regression coefficients can be tested by means of the loglikelihood ratio test.

The structural equation modeling approach to linkage analysis of multivariate phenotypes is a generalization of the univariate case:

$$\mathbf{P}(i,j) = \mathbf{\Lambda}_a \mathbf{A}(i,j) + \mathbf{\Lambda}_e \mathbf{E}(i,j) + \mathbf{\Lambda}_q \mathbf{Q}(i,j) + \mathbf{\Lambda}_d D(i,j) \qquad (6)$$

Here $\mathbf{P}(i,j)$ represents the (px1) random vector of phenotypic (centered) scores of sib j in sibship i. The $(p \times n_a)$ matrix $\mathbf{\Lambda}_a$ contains regression coefficients relating the p phenotypes to n_a latent additive genetic factors in the $n_a \times 1$ vector $\mathbf{A}(i,j)$. The matrices $\mathbf{\Lambda}_e$, $\mathbf{\Lambda}_q$, and $\mathbf{\Lambda}_d$ are defined in the same manner. Similarly, the vectors $\mathbf{E}(i,j)$ $(n_e \times 1)$, $\mathbf{Q}(i,j)$ $(n_q \times 1)$, and $\mathbf{D}(i,j)$ $(n_q \times 1)$, are vectors containing unshared environmental deviation scores, QTL additive deviation scores, and QTL dominance deviation scores. As above all the deviation scores have zero mean and are standardized. The partitioned (2px2p) covariance matrix, $\mathbf{\Sigma}_i$ of the multivariate phenotypic scores $\mathbf{P}(i,1)$ and $\mathbf{P}(i,2)$ equals:

$$\mathbf{\Sigma}_i = \begin{bmatrix} \mathbf{\Sigma}_{11i}\mathbf{\Sigma}_{21i}^t \\ \mathbf{\Sigma}_{21i}\mathbf{\Sigma}_{22i} \end{bmatrix}$$

where $\mathbf{\Sigma}_{11i} = \mathbf{\Sigma}_{22i}$. Assuming the latent variables $(\mathbf{A}, \mathbf{D}, \mathbf{E}, \mathbf{Q})$ are uncorrelated, the (pxp) covariance matrix $\mathbf{\Sigma}_{11i}$ equals:

$$\mathbf{\Sigma}_{11i} = \mathbf{\Lambda}_a\mathbf{\Lambda}_a^t + \mathbf{\Lambda}_e\mathbf{\Lambda}_e^t + \mathbf{\Lambda}_q\mathbf{\Lambda}_q^t + \mathbf{\Lambda}_d\mathbf{\Lambda}_d^t \qquad (7)$$

and the (pxp) cross covariance matrix $\mathbf{\Sigma}_{21i}$ equals:

$$\mathbf{\Sigma}_{21i} = \mathbf{\Lambda}_a\,[.5\otimes\mathbf{I}]\,\mathbf{\Lambda}_a + \mathbf{\Lambda}_q\,[\pi_i\otimes\mathbf{I}]\,\mathbf{\Lambda}_q + \mathbf{\Lambda}_d\,[p[1]_i\otimes\mathbf{I}]\,\mathbf{\Lambda}_d, \qquad (8)$$

where $\otimes$ is kronecker matrix multiplication and $\mathbf{I}$ is the identity matrix of appropriate dimension (the result of $[.5\otimes\mathbf{I}]$ is a diagonal matrix with .5 on the diagonal). Assuming the phenotypic data is approximately normally distributed, parameters in the matrices $\mathbf{\Lambda}_a$, $\mathbf{\Lambda}_q$, $\mathbf{\Lambda}_d$, and $\mathbf{\Lambda}_e$ can be estimated by maximizing the raw data loglikelihood function, and tests of significance based on the loglikelihood ratio test.

SIMPLIFYING ASSUMPTIONS AND MODEL PARAMETER VALUES

We assume that the QTL has 2 equi-frequent alleles and that its alleles are either codominant, or dominant. We assume that we have a marker situated zero cM away from the QTL, i.e. the QTL and the marker are adjacent on the chromosome. The marker has an infinite number of alleles (polymorphic information content, or PIC = 1), 16 alleles (PIC = .934), or 8 (PIC = .861). Regardless of the PIC value, the marker alleles are equi-frequent.

Under these simplifying assumptions, Table 17.2 in Haseman and Elston[9] can be used to derive a limited number of expected groups which are defined by different combinations of the values for π_i and $p[1]_i$. The number of sib-pairs within each group depends directly on the number of equi-frequent alleles at the marker locus, or, equivalently on the PIC value of the marker[20]. Depending on whether the

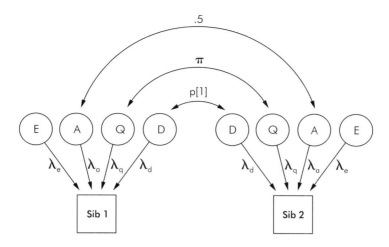

Figure 17.1 – Path diagram with observed phenotypes in sib 1 and sib 2 represented by
squares and latent variables E (individual-specific environment), A (additive genetic background), Q (additive QTL effects) and D (non-additive QTL effects) represented by
circles. The correlation between additive genetic background influences is 0.5, the correlation between additive QTL effects equals the proportion of alleles shared
identical-by-descent and the correlation between non-additive QTL effects is p[1]: the
probability that siblings share all alleles identical-by-descent.

number of markers, m, is infinite or not, and depending on whether the QTL is
codominant or dominant, Table 17.1 shows that we have 3 (when m=∞), 5 (when
PIC < 1 and there is no dominance), or 7 distinct groups (PIC < 1 and dominance
at the QTL locus). Because π_i and $p[1]_i$ only assume the 7 combinations shown in
Table 17.1, we can use multi-group covariance structure modeling to estimate
parameters.

For the power calculations we considered a scenario in which the QTL effect
explains 25% of the total variance in a trait. Background genetic influences also
account for 25% of the variance so that the total heritability of the trait is 50% and
the amount of variance explained by (random) environmental factors is also 50%
(actual values for variances used to construct the simulated covariance matrices
were Vq = λ_q^2 = 3 (for the non-additive QTL Vq =λ_q^2 = 2 and Vd = λ_d^2 = 1), Va
= λ_a^2 = 3 and Ve = λ_e^2 = 6. In the univariate covariance structure model we have:
λ_a =$\sqrt{3}$, λ_e =$\sqrt{6}$, and λ_q =$\sqrt{3}$, and λ_d =0, for the codominant QTL, or λ_q =$\sqrt{2}$,
and λ_d =$\sqrt{1}$, for the dominant QTL.

In the multivariate case, we have 4 phenotypes. The interrelationship between these phenotypes are determined by the matrices of regression coefficients. We introduce the following values in the case of a codominant QTL:

$$\Lambda_a^t = [\sqrt{3} \quad \sqrt{3} \quad \sqrt{3} \quad \sqrt{3}]$$

$$\Lambda_e = \begin{bmatrix} \sqrt{2} & \sqrt{4} & 0 & 0 & 0 \\ \sqrt{2} & 0 & \sqrt{4} & 0 & 0 \\ \sqrt{2} & 0 & 0 & \sqrt{4} & 0 \\ \sqrt{2} & 0 & 0 & 0 & \sqrt{4} \end{bmatrix}$$

$$\Lambda_q^t = [\sqrt{3} \quad \sqrt{3} \quad \sqrt{3} \quad \sqrt{3}]$$

In the case of a dominant QTL:

$$\Lambda_q^t = [\sqrt{2} \quad \sqrt{2} \quad \sqrt{2} \quad \sqrt{2}],$$

and

$$\Lambda_d^t = [\sqrt{1} \quad \sqrt{1} \quad \sqrt{1} \quad \sqrt{1}].$$

In the multivariate case, we assume that the 4 variables are influences by a single additive polygenic factor, and a single QTL. The unshared environmental influences are in part common to the 4 phenotypes, and in part specific to the 4 phenotypes. As mentioned in the Introduction, the model may arise when a phenotype is measured repeatedly over a short time span.

POWER CALCULATIONS

We refer to the true model, including the QTL factor, as H1, and we refer to the false model, excluding the QTL, as H0. Power equals $1-\beta$, where β is:

$$\beta = \text{prob(accepting H0 | H1 is true)},$$

i.e., the probability of a type II error. In the present context this means that there is a QTL effect, but that it is not detected. To calculate the power we follow the procedure described in Neale and Cardon[2] (1992, p. 190 ff.; see also 3.22). First we construct covariance matrices according to the true model, which includes the QTL. Next we use the Mx program[4] to fit the false model to these matrices using maximum likelihood (ML) estimation. The total number of sib-pairs, N, is chosen arbitrarily in fitting this model (say, 1000, or 5000). As is clear from Table 17.1, the distribution of this total N over the groups depends on the number of equi-frequent marker alleles and on the gene action of the QTL (codominant or dominant). The distribution of the goodness of fit index under the false model is distributed as a non-central chi-square variate. The exact form of the distribution

Table 17.1 – Distribution of π and p[1] given m, number of equi-frequent marker alleles, for m=8 (PIC=.861), m=16 (PIC=.934), m=32 (PIC=.968), and m= ∞ (PIC=1)*.

group	π	p[1]	frequency	m=8	m=16	m=32	m=[∞]
1	0	0	$\frac{1}{4}(m^3-2m^2+1)/m^3$	.1879	.2188	.2343	0.25
2	0.25	0	$(m-1)/m^2$	.1093	.0585	.0302	0
3	0.50	0	$\frac{1}{2}(m^2-2m+1)/m^2$	.3828	.4394	.4692	0.5
4	0.50	0.25	$1/m^2$	.0156	.0039	.0009	0
5	0.50	0.50	$\frac{1}{2}(m-1)/m^3$	.0068	.0018	.0004	0
6	0.75	0.50	$(m-1)/m^2$	.1093	.0585	.0302	0
7	1	1	$\frac{1}{4}(m^3-2m^2+1)/m^3$	.1879	.2188	.2343	0.25

* In the event of PIC = 1, we have three groups (1,3,7); in the event of PIC<1 and a codominant QTL, we have 5 groups (groups 3,4,5,are collapsed into a single group); in the event of PIC < 1 and a dominant (or recessive) QTL, we have 7 groups.

depends on the number of degrees of freedom, and the so-called non-centrality parameter (NCP). The number of degrees of freedom is simply the difference in the number of parameters between the true model (including the QTL) and the false model (without the QTL). The NCP equals the chi-square for the false model as reported by the program (i.e., Mx). Given N, the non-centrality parameter and the pre-specified α (e.g., .05, or .001), one can calculate the power to reject the false model, and one can calculate the required N to reject the false model, given a predetermined power. Conveniently, Mx carries out all the necessary calculations automatically. Below we report the required number of sib-pairs to reject the false hypothesis, given a power of .80 and an α of .001.

RESULTS AND DISCUSSION

Table 17.2 and 17.3 contain the number of sibling pairs needed to detect the effect of a QTL explaining 25% of the phenotypic variance. Table 17.2 summarizes the power calculations for a codominant QTL. As is expected (Fulker and Cherny, 1996), fitting the bivariate model gives better results in terms of power than analyzing squared difference scores. Regardless of PIC, the latter requires about a factor 1.35 more subjects than the former to achieve the same power. Multivariate model fitting involving all 4 phenotypes gives a substantial increase in power: 65% fewer subjects are required than in the univariate analysis. If the loadings of the QTL on the repeated measures of the phenotype can be constrained equal to each other, the increase in power is even larger, because the QTL effect can then be tested against 1 degree of freedom. The effect of PIC is as expected: the more informative the marker is, the more powerful the test of the QTL. Regardless of the test used, the reduction in the number of required sibpairs is about the same (from PIC=.93 to 1.0, about a factor .93). In terms of an ANOVA, one could say that PIC and 'type of test' both have a main effect on the required number of sibpairs, but that an interaction is absent.

Table 17.3 presents the number of sibling pairs required to detect the presence of a dominant QTL effect, the additive QTL component and the test of a QTL effect when dominance is ignored when fitting the full model (last 4 rows in Table 17.3). First, it is clear that the detection of the dominance variance of the QTL requires very large sample sizes. Multivariate modeling does substantially reduce the number of required sibpairs, but even the most powerful test still required over 16000 sibpairs. The power to detect the presence of the additive and dominance QTL variance simultaneously is much greater (second 4 rows in Table 17.3). Here the required samples sizes are comparable to those shown in Table 17.2. As it is very difficult to detect the dominance deviation, we finally investigate the power to detect the dominant QTL, under the circumstance that it is fit as a codominant QTL. This means that we model the QTL effect using a single parameter. The result (last 4 rows in Table 17.3) are very similar to those shown in Table 17.2. As in Table 17.2, there does not seem to be any interaction between the effects of PIC and the effects of 'type of test'.

The considerable increase in power associated with the multivariate test, suggest that it is advisable to collect multiple indicators of the phenotype under consideration or measure the phenotype repeatedly at multiple timepoints. An interesting question concerns the returns in terms of power of adding indicators. Figure 17.2 displays the required number of subjects to detect the codominant QTL when 1 to 9 indicators are analyzed. Again we consider the same three PIC values.

In Figure 17.2 we see that there is a dramatic increase in power when going from 1 to 2 and from 2 to 3 indicators. Beyond 3 indicators the increase in power is small, and beyond 5 indicators, the power actually decreases. Although the minimum number of required subjects is observed at 5 indicators, 3 or 4 indicators are sufficient. Needless to say, these particular results cannot be generalized to other parameter values, or genetic covariance structure models. However, it is very likely that the observed diminished returns will hold regardless of the details relating to the model.

In an earlier paper[8] we explored several strategies to analyze multivariate phenotypes. We found that when the multivariate information was summarized into a genetic factor score[23,24] no information was lost compared to fitting the complete multivariate model. This is a useful result because working with multivariate phenotypes may pose a problem in studies that selectively genotype extreme scoring sibling pairs. Multivariate selection of such pairs can be carried out on a genetic factor score which represents a subjects score on the latent genetic factor underlying the observations.

There are several ways to include a QTL in GCSM, which can be denoted the pi-

Table 17.2 – Number of sib-pairs to detect a codominant QTL with power=.80 and $\alpha=0.001$. The QTL accounts for 25%, background genes for 25% and environment for 50% of the total variance. For the multivariate data these effect sizes are the same for all 4 variables; environmental influences are split into variable specific effects (33%) and a common factor effect (17%).

analysis	PIC=1	PIC=.93	PIC=.86
Squared Difference score (df = 1)	2434	2598	2819
SEM Univariate (df = 1)	1795	1915	2077
SEM Multivariate (df = 4)	1155	1234	1340
SEM Multivariate (df = 1)	854	912	990

Table 17.3 – Number of sib-pairs to detect a QTL with equal allele frequencies, QTL additive effect = 16.6%, QTL non-additive effect = 8.3% (other effect sizes as in table 17.2; power=.80 and $\alpha=0.001$).

	PIC=1	PIC=.93	PIC=.86
Dominance effect			
Difference score (df = 1)	47,534	53,755	61,477
SEM Univariate (df = 1)	33,039	37,442	42,947
SEM Multivariate (df = 4)	22,052	24,928	28,522
SEM Multivariate (df = 1)	16,300	18,426	21,083
Dominance+Additive effect			
Difference score (df = 2)	2664	2861	3123
SEM Univariate (df = 2)	1961	2103	2291
SEM Multivariate (df = 8)	1324	1423	1554
SEM Multivariate (df = 2)	936	1005	1098
Total (D+A) effect			
Difference score (df = 1)	2432	2605	2836
SEM Univariate (df = 1)	1795	1920	2086
SEM Multivariate (df = 4)	1157	1240	1350
SEM Multivariate (df = 1)	855	916	998

hat approach and the IBD-distribution, or mixture, approach[6,7,25]. As it was more convenient for our present purposes, we have used the pi-hat approach in our power calculations. In unselected samples, these two approaches produce almost identical results.

In conclusion, on the basis of the present results, it appears that GCSM has more power than the original Haseman and Elston regression method[7] and that multivariate GCSM is more powerful than univariate GCSM[20].

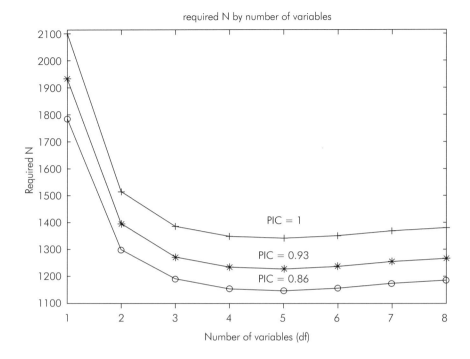

Figure 17.2 – number of sibpairs required to attain $\alpha=.001$ and $1-\beta=.80$ as a function of the number of phenotypic indicators (parameter values are the same as those in Table 17.1). The three plots (top to bottom) correspond to PIC=1, PIC=.93, and PIC=.86, respectively.

ACKNOWLEDGEMENT

We thank Mike Neale for helping with the model specification of a dominant QTL.

REFERENCES

1 Martin N.G. and Eaves L.J. (1977). The genetical analysis of covariance structure. *Heredity, 38*, 79–95, 1977.

2 Neale M.C. and Cardon L.R.(1992). *Methodology for Genetic Studies of Twins and Families* (NATO ASI Series D: Behavioural and Social Sciences-Vol. 67), Dordrecht: Kluwer Academic Publishers.

3 Jöreskog, K.G. and Sörbom, D. (1989). *LISREL 7: A guide to the program and applications* (2nd.). Chicago: SPSS press.

4 Neale M. (1997). *Statistical Modeling with Mx*, Department of Human Genetics, Box 3, MCV, Richmond VA 23298.

5 Boomsma D.I. and Molenaar P.C.M. (1986). Using LISREL to analyze genetic and environmental covariance structure, *Behavior Genetics*, *16*, 237–250.

6 Eaves L.J., Neale M. and Maes H. (1996). Multivariate multipoint linkage analysis of quantitative trait loci. *Behavior Genetics*, *26*, 519–525.

7 Fulker D.W., Cherny S.S., and Cardon L.R. (1995). Multipoint interval mapping of quantitative trait loci, using sib pairs. *American Journal of Human Genetics*, *56*, 1224–1233.

8 Martin N., Boomsma D., and Machin, G. (1997). A twin-pronged attack on complex traits. *Nature Genetics*, *17*, 387–391.

9 Haseman J.K. and Elston R.C. (1972). The investigation of linkage between a quantitative trait and a marker locus. *Behavior Genetics*, *2*, 3–19.

10 Goldgar D.E. (1990). Multipoint analysis of human quantitative genetic variation. *American Journal of Human Genetics*, *47*, 957–967.

11 Fulker D.W. and Cherny S.S. (1996). An improved multipoint sib-pair analysis of quantitative traits. *Behavior Genetics*, 26, 527–532.

12 Kruglyak L. and Lander E. (1995). Complete multipoint sib pair analysis of qualitative and quantitative traits. *American Journal of Human Genetics*,, 57, 439–454.

13 Carey G. and Williamson J.(1991). Linkage analysis of quantitative traits: Increased power by using selected samples. *American Journal Human Genetics*, *49*, 786–796, 1991.

14 Eaves L. and Meyer J. (1994). Locating human quantitative trait loci: Guidelines for the selection of sibling pairs for genotyping. *Behavior Genetics*, *24*, 443–455.

15 Cardon L.R. and Fulker D.W. (1994). The power of interval mapping of quantitative trait loci using selected sib pairs. *American Journal of Human Genetics*, *55*, 825–833.

16 Risch N. and Zhang H. (1995). Extreme discordant sib pairs for mapping quantitative trait loci in humans, *Science*, *268*, 1584–1589.

17 Gu C., Todorov A., and Rao D.C.(1996). Combining extremely concordant sibpairs with extremely disconcordant sibpairs provides a cost effective way to linkage analysis of quantitative trait loci, *Genetic Epidemiology*, 13, 513–533.

18 Dolan C.V. and Boomsma D.I. (1998). Optimal selection of sib-pairs from random samples for linkage analysis of a QTL using the EDAC test. *Behavior Genetics*, 28, 197–206.

19 Matsueda R.L. and Bielby W.T. (1986). Statistical power in covariance structure models. In: N. B. Tuma (Ed.). *Sociological Methodology, 1986*. San Fransisco: Jossey-Bass.

20 Boomsma D.I. and Dolan C.V. (1998). A comparison of power to detect a QTL in sib-pair data using multivariate phenotypes, mean phenotypes, and factor-scores, Behavior Genetics, 28, 329–340.

21 Sham P. (1998). *Statistics in Human Genetics*. New York: John Wiley & Sons.

22 Saris, W.E. and Satorra, A. (1993). Power Evaluations in Structural Equation Modeling. In: K.A. Bollen and J. S. Long, (Eds.). *Testing structural equation models*. p.181–204. Newbury Park: Sage Publications.

23 Boomsma D.I., Molenaar P.C.M., Orlebeke J.F.(1990). Estimation of individual genetic and environmental factor scores, *Genetic Epidemiology*, 7, 83–91.

24 Boomsma D.I., Molenaar P.C.M., Dolan C.V. (1991). Estimation of individual genetic and environmental profiles in longitudinal designs, *Behavior Genetics*, 21, 241–253.

25 Dolan C.V., Boomsma D.I., Neale MC, A simulation study of the effects of assigning prior IBD probabilities to unselected sib-pairs in covariance structure modeling of a QTL test, Am J Human Genetics, 64, 268–280, 1999.

18

QTL MAPPING WITH SIB-PAIRS: THE FLEXIBILITY OF Mx

Michael C. Neale

ABSTRACT

The search for quantitative trait loci (QTLs) for complex traits is particularly difficult when the measurement of the traits is problematic. In this case combining the best available psychometric methods with methods for detecting QTL can offer considerable advantages in statistical power over other approaches. Several models for comorbid disorders and multivariate data are described, and methods for testing for single and multilocus effects are presented.

The main aims of this chapter are to describe the advantages of structural equation modeling (SEM) and related approaches to the analysis of covariance structure in the search for QTLs, and the quantification of their effects. The chapter begins with a brief introduction to the biometrical model behind the approach, including the concepts of identity by descent, the predicted correlation between relatives, and finite mixture distributions. There follows a more detailed 'how-to' section on the structural equation modeling of sib-pair data using Mx with two main approaches. Liberal use of graphics for path diagrams that define the models is intended to make the treatment accessible to readers from a variety of backgrounds. Following the description of the basic model, a number of extensions to the approach are discussed, most of which may be implemented with the current version of Mx (1.47).

MODELING THE EFFECTS OF QTLS

Allelic effects on quantitative traits

Our starting point for modeling the effects of an allele on a quantitative trait is the elegant description by Ronald Fisher in his classic 1918 paper "The correlation between relatives on the supposition of Mendelian inheritance". The model considers a diallelic locus with possible genotypes A_1A_1, A_1A_2 and A_2A_2 as shown in Figure 18.1. It is a very simple model, especially in the respect that only diallelic loci are considered, because in man most functional polymorphisms involve more than two variants[1]. Nevertheless it is a good starting point because often variants can be grouped according to their effects on a trait such that the locus operates according to diallelic principles.

If allele A_1 has frequency p in the population, and allele A_2 has frequency $q = 1 - p$, simple algebra may be used to derive the effects of the locus on population variation for the trait. The algebra involves computing the population mean, weighting the squared deviations of each genotype from the mean by its frequency in the population, and summing over the three genotypes. It can be shown that the variance due to the locus, in terms of additive and dominance deviations, is

$$V_G = 2pq \, [a - d(p - q)]^2 + 4p^2q^2d^2$$

It may further be shown that this is the sum of the additive genetic variance $V_A = 2pq[a - d(p - q)]^2$ and the dominance genetic variance $V_D = 4p^2q^2d^2$. In a classical polygenic model with n independent loci affecting a trait, the additive and dominance genetic effects are the simple sums of these quantities across the n loci. When n is large and each locus is of small and equal effect, the central limit theorem predicts that a normal distribution will emerge. However, if one locus has a large effect the population would consist of a mixture of three normal distributions with mean deviations $-a$, d, and a (the genotypic means).

Identity by descent

A fundamental concept in the analysis of the effects of QTL is whether a pair of relatives shares zero, one, or two alleles *identical by descent* (IBD) at a locus. This idea should not be confused with *identity by state* which merely indicates whether two alleles are physically the same, e.g., A_1 and A_1. Two alleles are defined as IBD "if they are derived without mutation by transmission along a common pathway from the same gene in a common ancestor"[5]. For statistical purposes, what is relevant is whether the alleles were inherited from the same chromosome of a parent, because we are interested in whether any alleles in that region give rise to variation in the trait being studied.

In the case of full siblings, it is easy to tabulate the possible IBD outcomes by

Mean	$-a$	d	$+a$
Genotype	A1A1	A1A2	A2A2
Frequency	p^2	$2pq$	q

Figure 18.1.– Mean values of possible genotypes at a diallelic locus, in terms of deviations due to additive (a) and dominance (d) effects.

writing parents' alleles as *AB* for the father and *CD* for the mother. Their possible offspring are *AC, AD, BC* and *BD,* and the possible pairwise combinations of these offspring are shown in Table 18.1[5,6]. The cells of this Table indicate the number of alleles shared IBD by each of the 16 possible sib pair types. Since each combination is expected to be equally frequent, the expectation is that one-fourth of the pairs will be IBD 2, one half will be IBD 1, and one-fourth will be IBD 0).

During gamete formation, a process known as recombination causes each gamete to be formed from a mixture of the two parental chromosomes. This crossing-over or forming of *chiasmata* in humans occurs approximately once per Morgan of map length. Two loci close together on the genome are unlikely to have a recombination event between them; they are said to be closely linked. As the distance increases, so the likelihood of one or more recombinations increases to the point where there is a .5 probability that the two loci were inherited from the same parental chromosome. Computer programs such as Mapmaker/sibs[7] make use of this probabilistic model and data collected on marker loci to compute the probability that a sibling pair shares zero, one or two alleles IBD for any location on the genome. The closer the markers are to this location, and the more informative they are, the more precisely the IBD status can be discerned. With highly informative markers close to the locus of interest, it is often possible to obtain a set of probabilities such as 1,0,0 for p(IBD0), p (IBD1), p(IBD2), indicating that the IBD status is known.

Another valuable use of the information in Table 18.1 is to compute the predicted co-variances between additive genetic and dominance genetic components of pairs of relatives. The dominance covariance is simply *p(IBD2)* which is 4/16 = .25 for sibling pairs. The additive genetic covariance is computed as *p(IBD2)* + .5*p(IBD1)*, or 4/16 + 1/2 × 8/16 = 1/2 for sibling pairs. For MZ twins, *p(IBD2)* = 1 so both the additive and the dominance components correlate perfectly.

Correlation between relatives

Fisher's 1918 paper provided the basis for modeling the resemblance between rela-

Table 18.1: Number of alleles shared identical by descent for a pair of full-siblings. Parental genotypes are AB and CD.

	AC	AD	BC	BD
AC	2	1	1	0
AD	1	2	0	1
BC	1	0	2	1
BD	0	1	1	2

tives under the assumption of polygenic inheritance. The same basic model is used today in the routine analysis of twin and family studies. Fulker[8] was one of the first to begin the systematic application of the methods to monozygotic (MZ) and dizygotic (DZ) twins for the purposes of examining the genetics of behavior. Eaves[9] and others continued to develop extensions of these methods in the 1970's and 80's and built the methodological framework that is the core of the modern methods. In the past 10 years structural equation modeling has become the method of choice for most genetic epidemiological studies of twins and families[4], and therefore it is a good foundation for the modeling of the effects of specific loci in addition to background polygenic or oligogenic factors.

Probably the most commonly used model for familial resemblance is the 'ACE' model, which is identified in research designs consisting of MZ and DZ twins (the 'classical twin study'), and certain types of adoption study. In this model, A stands for Additive genetic, C stands for Common environment, and E stands for random Environment not shared between relatives. The variance due to additive genetic factors has already been described above; it will make MZ twins covary to twice the extent of DZ twins, assuming random mating. Common environment is defined as those environmental factors shared by pairs of twins or siblings that have the same effects on relatives reared together, i.e., those factors that cause relatives to be alike. Conversely, random environmental factors are by definition those that do not make members of the same family resemble one another. Estimates of the effects of the random environment factor will usually include measurement error, and often include $G \times E$ interaction.

STRUCTURAL EQUATION MODELING OF QTLS

The advantages of structural equation modeling approaches have been described in detail elsewhere, for both genetic[4] and non-genetic applications[10-12]. Their valuable properties include: assessment of overall fit of a model; testing of specific hypotheses about causation and correlation; and the availability of confidence intervals on parameters of interest. Much modern model fitting is based on likelihood theory. Maximum likelihood parameter estimates have numerous desirable statistical properties, including that they are invariant to transformation, are asymptotically unbiased, and have minimum variance of all asymptotically unbiased estimators. Sometimes structural equation models are criticized for being too narrow, although this is difficult to understand when so much of modern statistics is subsumed in this framework, including: analysis of variance; multiple regression; factor analysis; principle components analysis; discriminant analysis; canonical correlations; hierarchical linear modeling; growth curves; and dynamical systems[12-15]. Quantitative genetic models have an especially strong rationale due to the clear mathematical treatment of the transmission of genetic factors. It seems no accident that path analysis (a.k.a. structural equation modeling with diagrams) was invented by a geneticist, Sewall Wright[16].

Two main types of variable are permitted in path diagrams: observed, shown in boxes, and latent (not observed) shown in circles. Relations between these variables may be either causal (single-headed arrows) or correlational (double-headed arrows) in accordance with the investigator's theory. It must be noted that not all models are distinct, in that several different configurations of causal or correlational paths can predict a set of covariances equally well. It is a matter for good study design to ensure that the models are identified and make conceptual sense.

Structural equation models for the resemblance of three classes of relatives are shown in Figure 18.2. As noted above, MZ twins share all their genetic material; they have $p(IBD2) = 1$ for all loci, so their additive and dominance genetic components A and D correlate perfectly. Environmental effects are partitioned into two sources: those that are common to the members of a twin pair (C), and those that are unique to each individual (E).

DZ twins are assumed to share environmental influences to the same extent as MZ twins, which is the 'equal environments' assumption of twin studies. This assumption has been tested empirically for many cognitive and psychiatric variables[17-19] and almost invariably has been found to be supported. The same assumption may be made for non-twin siblings and for unrelated individuals (adoptees) reared in the same household, though again, especially with age differences between these relatives, the assumption should be tested.

To identify all four parameters of the models in Figure 18.2, it is necessary to analyze data from all three groups simultaneously. In the absence of data collected from unrelated pairs, i.e., the classical twin study, either C or D must be assumed to be absent. With data from only sibling pairs or DZ twins, A and C or D are confounded, so while familial resemblance may be detected (Figure 18.2ii), its possible genetic or environmental origin cannot be determined. From these basic models we can now turn to the addition of data on putative QTLs, and the estimation of the effects of a locus on the phenotype, in addition to the residual polygenic (r) and environmental components of variation.

Fully informative markers.

With sufficiently informative markers and a dense marker map, most sibling pairs can be classified as sharing zero, one or two alleles IBD at a putative location. Such a classification provides a direct parallel with the multiple group structural equation model described in the preceding section. Sibling pairs that share both alleles IBD are just like MZ twins at that locus; their additive QTL and their dominance QTL components correlate perfectly. However, while sibling pairs that share one allele IBD will share half the additive genetic effects of the locus, they have $p(IBD2) = 0$ at that locus and therefore do not share any dominance genetic effects from the locus. Finally, sibling pairs that do not share any alleles IBD will

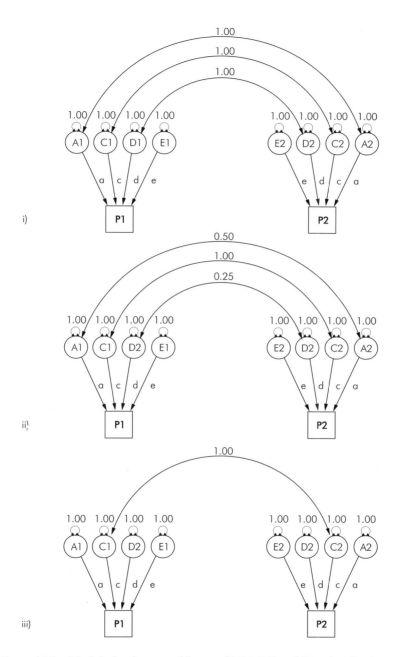

Figure 18.2 – Models for the resemblance of MZ, DZ and Unrelated pairs reared in the same family. Additive genetic (A), dominance genetic (D), common environment (C) and random environment (E) components cause variation in the phenotypes of a pair of siblings (P1 and P2).

not correlate for either additive or dominance effects of the QTL and therefore parallel unrelated sibling pairs. The statistical power of models of this type was considered by Nance & Neale[20]. With random samples, power is generally quite low, even when only additive genetic variation at the locus is considered. The power is low because the information is in the IBD2 vs. IBD0 contrast, but these pairs only constitute half of a random sample of pairs which occur at rates of .25 IBD2, .5 IBD1 and .25IBD0 in the population.

The $\hat{\pi}$ ('pi-hat') method.

When markers are less than fully informative, and when they are not located close to the putative QTL of interest, it is not possible to classify sibling pairs into IBD groups unambiguously. Instead, we may summarize the marker information as f_i, the set of three IBD probabilities:

$$f_i = p(IBD2), p(IBD1), p(iBD0)$$

and use them to compute $\hat{\pi}$, an estimate of the correlation between the additive effects of the QTL:

$$\hat{\pi} = p(IBD2) + .5p(IBD1).$$

The parallel statistic for the correlation between the dominance effects of the QTL would simply be $p(IBD2)$, but this is not commonly used. One reason for this omission is that there is generally lower statistical power to detect dominance variance; a second is that there is usually additive genetic variation at any locus that displays dominance, so for the purposes of establishing linkage the additive QTL model is usually sufficient.

The models described in previous sections may be implemented with any structural equation modeling program that has multiple group capabilities, but a model that involves $\hat{\pi}$ requires some special features. Most important, $\hat{\pi}$ may be different for each pair of siblings in the sample, which would require as many groups as there are pairs in the sample. This would clearly be an inefficient approach to programming such a problem! In addition, there would only be one sibling pair per group which would mean that the use of covariance matrices to summarize the data would be inappropriate as there would be zero variance. We can overcome this second limitation by maximizing the likelihood of the raw data vectors themselves, using the log-likelihood function:

$$\log L_M = \log |2\pi\Sigma_i|^{-n/2} - .5(x_i - \mu)'\Sigma_i^{-1}(x_i - \mu)$$

where Σ_i is the covariance matrix corresponding to the ith sib pair, μ is the (column) vector of means of the variables, n is the number of variables in the pair, and $|\Sigma_i|$ and Σ_i^{-1} denote the determinant and inverse of the matrix Σ_i, respectively. The usual formula for normal theory log-likelihood relies on a single population covariance matrix Σ and no subscript is required. In the present case, the predicted covariance matrix depends on the estimate $\hat{\pi}_i$:

$$\Sigma_i = \begin{pmatrix} q^2 + r^2 + e^2 & \hat{\pi}_i\, q^2 + r^2 \\ \hat{\pi}_i\, q^2 + r^2 & q^2 + r^2 + e^2 \end{pmatrix}$$

which can differ for every sib pair. The term r refers to residual additive genetic variance not accounted for by the QTL, together with the effects of the shared environment.

It is possible to specify models of this type using the 'definition variables' feature of package Mx[21]. For didactic purposes it is also useful to specify the model using the path diagram drawing feature of the Mx graphical interface. Figure 18.3 shows a path diagram of this model. The diagram is a mathematically complete description of the model and therefore can be used to fit the model to the data, after which the results are automatically displayed on the diagram. This diagram, its associated data and output files and the Mx software are available for free download at the website http://www.vipbg.vcu.edu/mx/cambridge.

(1)

A delightful advantage of the $\hat{\pi}$ method is that it generalizes very easily to larger sibships and pedigrees. For parents and offspring $\hat{\pi}$ is one half (because $p(IBD1) = 1$). It is also possible to enter a specific $\hat{\pi}_{ijk}$ for the covariance between sibs j and k in sibship i.

Models that change for every case in the sample can be somewhat computationally intensive, especially with large samples. With N cases it is necessary to compute and invert N predicted covariance matrices in order to evaluate Equation 1. This process must be repeated for every set of trial values used during the search for the maximum likelihood solution. Some advantage is to be gained by sorting the data by $\hat{\pi}_i$, as Mx will recognise that the model has not changed from one case to the next and will not re-evaluate the model. To some extent this computational burden may be alleviated by taking an alternative approach that uses weighted likelihoods, to which we now turn.

The weighted likelihood method.

Specifying the covariance between the additive effects of the QTL of sibling pair i as $\hat{\pi}_i$ is an effective and flexible approach[22] but it has its limitations. The reader will probably have noticed that a specific value of $\hat{\pi}$ could be obtained from a variety of sets of IBD probabilities f_i. For example, $\hat{\pi} = .5$ is consistent with both $f_i = [.4, .2, .3]$ and $f_i = [.2, .6, .2]$ yet these two sets contain quite distinct information about the IBD status of the pair. Therefore, $\hat{\pi}$ is not a sufficient statistic to describe what is known about the IBD status.

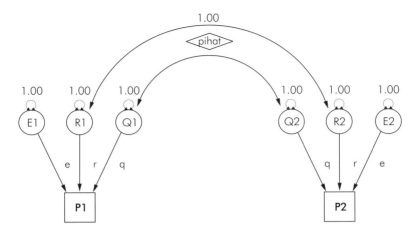

Figure 18.3 – Model for resemblance between siblings as a function of estimated sibling covariance, $\hat{\pi}$, for a putative QTL

The reality of the biological situation is that any particular sib pair does not correlate $\hat{\pi}$ at a QTL. Pairs belong to one of three IBD classes, but the marker data do not precisely identify which one. A natural mathematical representation of this situation is to use a finite mixture distribution[23], so called because it is formed by a mixture of a finite number of distributions. The log-likelihood of a mixture of r distributions is given by:

$$L_M = \log \sum_{d\,=\,1}^{r} w_d\, \phi_d(x)$$

where $\phi_d(x)$ is the density function for distribution d and w_d is the weight, or mixing proportion, of that distribution. In our case we can use the IBD probabilities f_i as the weights and compute covariance matrices according to Equation 1 with $\hat{\pi}$ set at one, point five, and zero. These predicted covariance matrices are invariant with respect to the sib pair in question and therefore, regardless of sample size, only three covariance matrices need to be inverted per set of parameter values being evaluated in the search for the optimum.

Statistically, the mixture distribution method seems superior. It is a much better representation of the true situation and it makes full use of the IBD probabilities instead of a summary $\hat{\pi}$ statistic. In both simulations and applications to real data we have found increased statistical power. However, for small QTL effects the differences seem negligible[24].

Although it is computationally efficient and statistically powerful for large samples of sib pairs, one disadvantage of the mixture approach is that it does not generalize easily to larger sibships or to more complex pedigrees. While sibling pairs have three possible IBD configurations, this number rapidly increases with sibship size. For each pedigree in the sample it will be necessary to compute the probability of every possible IBD configuration of the sibship. This is not a simple exponential increase of the form 3^{m-1} with sibship size m because many configurations are not possible (for example if sibs 1 and 2 are IBD 2 they must share the same number of alleles IBD with any other sibling in the family). The number of configurations works out to be $2^{2m-3} + 2^{m-1} - 2^{m-2}$, which still climbs rapidly (3,10,36,136,528). The predicted covariance matrix will need to be computed for each of these configurations, and weighted by the appropriate probability.

At present it is not possible to specify models for mixture distributions with the Mx graphical interface, but it is possible to do it with the script language. A script is available on the website http://www.vipbg.vcu.edu/mx/cambridge.

Intrapair Differences

An interesting finding reported in the Fulker & Cherny[24] paper was that the power to detect linkage is substantially increased by using sib pair covariances instead of intrapair differences. This result is not especially surprising from the point of view of the information available; intrapair differences are not a sufficient statistic to describe the phenotypes of both members of a sibling pair. It might seem that the use of intrapair differences has nothing to recommend it, but this impression could be false.

It is possible that the variance of intrapair differences is more robust to violations of bivariate normality than the covariance of sib pair scores. The degree to which this advantage outweighs the loss of information is a question that could be answered by simulation. A further option may be to work with intrapair differences *and sums* which would recover the information lost, but which may still be more robust to violations of normality. In general it is better to identify the source of non-normality and control for it. Possible sources include non-random sampling, inherent scaling problems, and the presence of a major gene of large effect, among others. Appropriate model development can readily incorporate these factors, but there is much work still to be done in this area[25].

In a structural equation model it is simple to change from the model for variances and covariances of sibling phenotypes shown in Figure 18.3 to a model for intrapair differences, which is shown in Figure 18.4a. The previously observed phenotypes are now latent variables, and the observed difference variable is simply 1 times the sibling 1 phenotype and −1 times the sibling 2 phenotype. Figure 18.4b shows an extension of this model of differences to include intrapair sums as well as differences. Equivalent mixture distribution models can be developed in the same way.

EXTENSIONS TO THE SINGLE LOCUS UNIVARIATE MODEL

In this section I describe a number of ways in which the univariate models can be developed to cater for more than one putative QTL, and for different measurement strategies. At this point not all such models have been implemented and tested, but the basic principles are worth considering.

Two locus model

One straightforward extension to the model for sibling pairs is to obtain IBD probabilities for a second locus and estimate the effects of two QTLs simultaneously. Eaves et al [26] described a model of this sort for two unlinked loci and that will be considered here. Using the $\hat{\pi}$ approach, the two putative QTIs are simply modeled as two separate, uncorrelated latent variables, as shown in Figure 18.5. The $\hat{\pi}_A$ and $\hat{\pi}_B$ are computed exactly as for the univariate case but at two different locations.

The mixture distribution approach requires more careful construction. Sibling pairs may share zero, one or two alleles at either locus. Taken together, there are nine possible combinations of pair IBD status at the two loci, so the likelihood is written as a mixture of nine models. As long as the loci are unlinked, the vector of IBD probabilities for location one (f_i) and that for location two (g_i) may be multiplied $f_i' g_i$ to yield the nine different iBD probabilities. This nine-model mixture distribution is shown in Figure 18.6. It is colloquially referred to as the 'space invader' model because it resembles the ancient computer game. The model works very well with simulated data, producing parameter estimates that are very close to the true population values.

An interesting feature of the mixture distribution model specification is that it can easily be extended to allow for epistatic interactions between loci. For example, additive × additive interaction between two loci will increase the resemblance of pairs of relatives that share 1 or 2 alleles IBD at both loci. Pairs that are IBD 2:2 will covary a full V_{AA} component while those that are IBD 2:1 will share one-half V_{AA} and those that are IBD 1:1 will share one-fourth V_{AA}. Only those pairs that are IBD 2:2 will share a dominance × dominance interaction component.

Some modification is required to fit models in which the two loci are linked. As long as there is linkage equilibrium between the two loci, which is likely to be the case unless linkage is very tight – perhaps 1cM or less – it is reasonable to ignore linkage for the $\hat{\pi}$ method. For the mixture distribution approach we need to compute an appropriate set of joint probabilities. This can be done by using the marginal IBD probabilities from the two loci, f_{1i} and f_{2i}, forming diagonal matrices (F_{1i}) and (F_{2i}) from these vectors, and computing the matrix quadratic $(F_{1i})\mathbf{GRG}(F_{2i})$, where

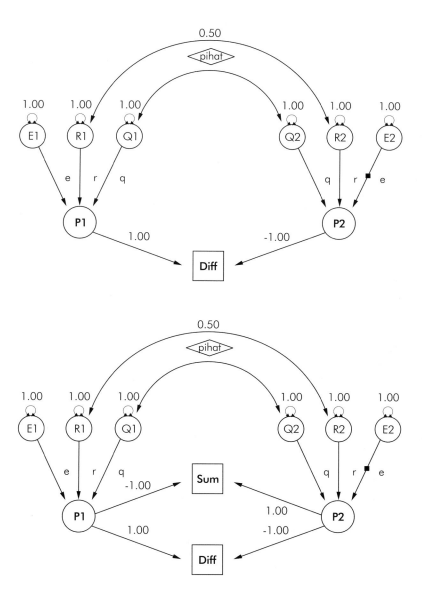

Figure 18.4 – Model for differences between siblings phenotypes (P1 and P2) as a function of estimated sibling covariance, $\hat{\pi}$, for a putative QTL. b) Model for sibling sums and differences.

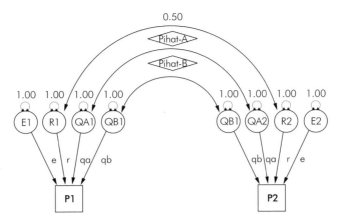

Figure 18.5 – Two locus model for sibling pairs, with QTL factors (Q_A and Q_B), residual familial resemblance (R) and random environment (E) components which cause variation in the phenotypes of a pair of siblings (P1 and P2). Marker data and multipoint methods are used to compute IBD sharing coefficients $\hat{\pi}_A$ and $\hat{\pi}_B$ at the two loci.

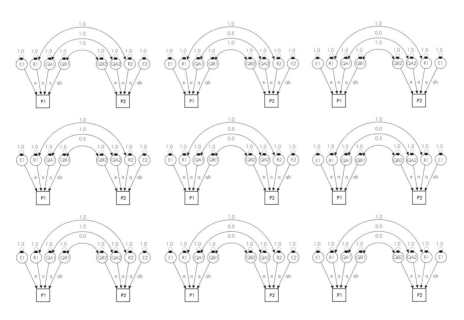

Figure 18.6 – Two locus mixture model for sibling pairs, with QTL factors Q_A and Q_B, residual familial resemblance (R) and random environment (E) components which cause variation in the phenotypes of a pair of siblings (P1 and P2). Marker data and multipoint methods are used to compute IBD sharing probabilities at the two loci which are used to weight the likelihoods for these nine possible models.

$$R = \begin{bmatrix} \Psi^2 & 2\Psi(1-\Psi) & (1-\Psi)^2 \\ 2\Psi(1-\Psi) & 2(1-2\Psi+2\Psi^2) & 2\Psi(1-\Psi) \\ (1-\Psi)^2 & 2\Psi(1-\Psi) & \Psi^2 \end{bmatrix}$$

and $\Psi = \theta^2 + (1-\theta)^2$ (Haseman & Elston, 1972). Multiplication by the diagonal matrix **G** with diagonal elements 4,2,4 is necessary to obtain appropriate marginal probabilities from the product with matrix **R**. The practical upshot of this result is that the nine IBD joint probabilities for the different IBD configurations of the sib pairs could be computed prior to analysis and used in place of the six IBD probabilities. Since **R** is symmetric, only six of these nine elements are unique, so only six definition variables would be required.

Multivariate models

A valuable feature of structural equation models is that they are very easy to extend to the multivariate case. Vogler[27] developed a matrix-based path analysis that is especially helpful for modeling multivariate familial resemblance. Being matrix-based it is straightforward to implement in Mx; indeed, Vogler's useful paper was part of the inspiration for Mx's development. A multivariate model of genetic and cultural transmission that uses data from MZ and DZ twins and their parents was developed using an early version of Mx[28]. More recently, a complex model for additive and dominant gene action, assortative mating, cultural transmission and special twin environments has been presented[29]. This model is identified with data from twins, their parents, their spouses and their children – a three-generation twin-family design.

Regardless of the type of family structure, be it simply pairs of one type of relative or the 80+ types of relationship in the three-generation twin-family design, it is possible to test a variety of explanations for the covariation between different traits. When data are collected from unrelated individuals, the study of covariation using structural equation modeling is called confirmatory factor analysis. These factor models may be implemented in a parallel form when there are multivariate data on relatives; such models are termed 'psychometric' or 'common pathway'. However, it is also possible to factor each separate component of variance – additive genetic, common and random environment – independently of one another. This type of model is termed 'biometric' or 'independent pathway'[4,30,31]. These models form the basis of most current multivariate analysis of data collected from genetically informative designs. Other models are described in the section on comorbidity below.

Multiple raters

A common approach to the assessment of behavior and behavioral problems in children is to use ratings made by parents or teachers. Naturally there is less than perfect agreement between raters about the behaviour of children; often the correlations between mother and father ratings are .6 or less and those between parents and teachers are .3 or less[32]. This lack of agreement could be due to several factors including: (i) the behavior of the child differs according to the parent that they are with, or the situation (home vs. school) that they are in; (ii) parents and teachers assess behavior inaccurately due to ignorance of the true behavior of the child; and (iii) parents and teachers are biased in their ratings. There may also be special contrast effects when parents are rating twins such that differences between the twins are exaggerated in the reports. All of these processes lead to decreased precision from any single rater. A latent variable model, as shown in Figure 18.7 allows partition of these sources with a concomitant increase in statistical power to detect the effects of a QTL acting on the 'true' component of behavior.

Comorbidity

Until recently, there was no good theoretical framework for the origin of comorbidity[33], and no empirical way to distinguish between alternative verbal hypotheses about its origin. Using Mx, we developed a variety of models for comorbidity to data on lifetime diagnoses of Generalized Anxiety Disorder and Major Depression[34]. These models address various configurations of *dimensionality, heterogeneity* and *causation* which might underlie the high rate of comorbidity for these disorders (within person tetrachoric correlation $r = .65$). A key theoretical issue is whether the pathology of one disorder directly increases risk for the second disorder, or if high liability (with or without the disorder) confers increased risk. This issue may be resolved with appropriate modeling of the expected frequencies of relatives of different types. To this end, we classified the following types of comorbidity:

- *Chance*: disorders are independent.
- *Alternate forms*: one underlying dimension of liability with one threshold and a stochastic outcome probability p of having one disorder and probability r of having the second.
- *Random multiformity*: two independent liability dimensions with one threshold per dimension. Those above threshold on dimension A have probability 1 of disorder A, and probability p of disorder B. A similar process operates with dimension B, but with probability r of giving rise to disorder A. It is expression of one disorder *per se* that gives rise to increased risk of the second disorder. The 'gateway' theory of substance abuse may be considered to be a form of multiformity model.
- *Extreme multiformity*: which is similar to random multiformity, but the increase in liability to the second disorder only occurs among those above a second, higher threshold.
- *Three disorders*: excess comorbidity (above chance rates) arises from a third independent liability distribution.

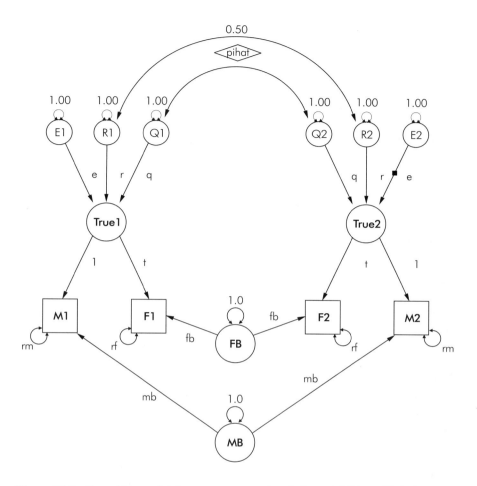

Figure 18.7 – Rater bias model for two parents rating their two children. Variation in the ratings is a linear combination of the ratee's true score *T*, parental rater bias *B*, and error *E*. Correlations between the ratings of child 1 by their mother (*M1*) and father (*F1*), and of child 2 (*M2*, *F2*) identify the model. *FB* represents father's rater bias, which causes his ratings of his children to correlate; *MB* is the rater bias due to the mother.

- *Correlated liability*: a direct analogue of standard bivariate models for correlated genetic, shared environmental, and specific environmental factors in twins[4].
- *Causal models*: implementation of direction of causation models [4,35-38]. Causation occurs at the level of the risk factors or the latent liability dimension, and is thus quite different from the multiformity models.

Although these models were developed in the context of the classical twin study, they are entirely appropriate for studies of sib pairs and studies of the effects of QTLs. To operationalize them for QTLs requires replacing the additive genetic components A in the models with QTL effects Q, and computing a weighted sum of the likelihoods according to the IBD probabilities of the sib pair. Doing so should yield additional power to detect QTLs and may provide a better explanation of observed patterns of comorbidity in the population.

Repeated measures

A second approach to improving the phenotypic 'signal' of a disorder is to assess subjects on different occasions. It is clear that test-retest reliability is modest for many psychiatric and medical diagnoses[39-41] and that repeated measures can reduce the unreliability[42,43]. Such methods have yielded substantially increased estimates of heritability of the liability to lifetime major depression[44]. In addition, we have shown that for test-retest reliabilities below .7 it is almost always more powerful to measure subjects twice than to double the sample size[45]. Such increased heritability – and the increased power to detect it – directly translates into increased power to detect QTLs. This increased power seems to exceed that obtained with other methods of combining information from multiple occasions, such as summing scores across occasions or using Boolean rules.

Figure 18.8b shows the sample sizes required to detect linkage to a qualitative trait using a random sample of sibling pairs. Data were simulated according to the model shown in Figure 18.8a which yields a QTL effect of 17.5% of the variance at a single occasion. It was further assumed that IBD status was known without error for these data, such as may be commonplace when SNP technology becomes available. The threshold was set to yield 25% of the population affected with the disorder. Three univariate diagnostic strategies were compared: measurement on one occasion only; affected at either occasion; and affected at both occasions. A fourth, bivariate analysis of both occasions was also tested. The bivariate analysis is clearly more powerful, requiring, for example, 3654 instead of 4541 pairs to achieve 80% power. These results for a test-retest correlation of .7 would be more impressive with lower correlations, which we know to be the case for several psychiatric and medical disorders[39-41,46]. For example, in our adult female twin sample we found that Major Depression may be a highly heritable disorder with low reliability[46]. This establishes a strong case for SEM analysis of retest measures in genetic studies.

Multiple relatives

While most of the discussion in this chapter has focussed on data collected from pairs of relatives, it is clear that considerable increases in statistical power may be obtained when larger pedigrees are available[47]. As Blangero and others have shown[48], it is relatively straightforward to employ $\hat{\pi}$ estimates computed for all possible pairs in a general pedigree. We have achieved the same results using some simple matrix algebra within the Mx program. The covariance between n_s relatives on n_t traits may be written using the Kronecker product ($\otimes$) as:

$$\mathbf{U} \otimes (\mathbf{CC'}) + \mathbf{I} \otimes (\mathbf{EE'}) + \mathbf{S} \otimes (\mathbf{QQ'})$$

where the matrices are defined as: $\mathbf{I}(n_s \times n_s)$, identity; $\mathbf{U}(n_s \times n_s)$, unit; $\mathbf{S}(n_s \times n_s)$ standardized (unit on diagonal) matrix with off diagonal elements containing $\hat{\pi}_{jk}$ for relative j with relative k; $\mathbf{C}$ ($n_t \times n_t$) lower triangular matrix to estimate residual relative covariances within and between traits; $\mathbf{E}(n_t \times n_t)$ a lower triangular matrix to estimate within-person variation and covariation due to factors not shared by relatives; and $\mathbf{Q}(n_t \times 1)$ a vector that contains the effects of the locus on the n_t measured traits.

Much more complicated is the weighted likelihood method. An array of possible IBD states for the relatives must be generated together with the probability (based on multipoint analysis of the genetic marker data) that this state exists in this pedigree. The number of possible states could be very large for even modest-sized pedigrees, so this method will be computationally intensive with large pedigrees. With current computer hardware limitations, it would not seem viable for multi-variate analyses of large samples of large pedigrees.

Selected samples

Two primary approaches to the analysis of selected samples appear suitable, and both can be implemented in the current version of Mx. First is to correct the likelihood by dividing it by the probability that a pedigree is expected to be ascertained. Doing so normalizes the likelihood to an overall probability of unity. While straightforward, this approach does have to make use of parameters of the model to compute the ascertainment correction, for it will depend on the correlation between relatives. In a mixture distribution approach, this correlation will vary as a function of the IBD state under examination. For the $\hat{\pi}$ approach the ascertainment correction will vary as a function of $\hat{\pi}$ and therefore may be computationally intensive to compute. Under normal theory threshold modeling the ascertainment correction will typically involve an n_s-dimensional integral for pedigrees of size n_s. The correction needs to be computed for each pedigree of distinct type for each iteration in the search for a solution, because the parameters of the model dictate the size of the ascertainment correction. This is one area in which the mixture distribution approach would seem to have a computational advantage.

The second main approach is to make use of non-random sampling theory as

a)

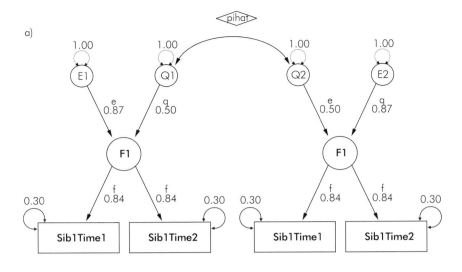

b)

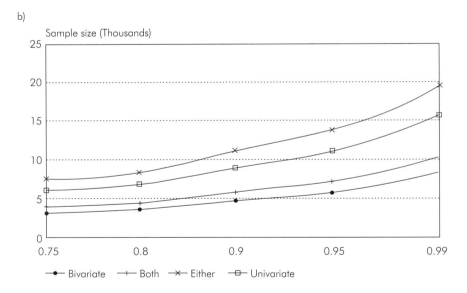

Figure 18.8 – (a) Path model for retest diagnostic data. Covariation between occasions is $f^2 = .7$ and is caused .25 by a QTL and .75 by random environmental factors. (b) Sample size required to detect linkage using different strategies for analyzing bivariate data simulated under the model in (a). The full bivariate analysis shows greater power than summarizing the data using an either or both diagnostic requirement.

described by Little & Rubin[49]. When a sample has been selected from a population by means of a screening instrument, it is possible to conduct a joint analysis of the screen and the phenotype of interest. Under these circumstances, maximum likelihood estimation should yield unbiased parameter estimates[26] although appropriate modeling of the relationship between the screen and the target phenotype of interest is required.

ASSOCIATION AND HAPLOTYPE ANALYSES

While linkage, which is robust, may be the mainstay of genetic analyses for several years to come, association and haplotype-based analyses will most likely continue to grow in importance. One natural sequence is to establish evidence for linkage in a particular region, and then to obtain a large number of closely linked markers in that region to be used for haplotype scanning[50]. With the advent of SNP technology and gene chips[51], haplotype scanning procedures will become more viable as an initial approach. While haplotype scanning has the potential advantage of being able to identify a much narrower region (1cM or less) it is not without drawbacks. A major problem is that it relies on certain population genetics assumptions, and is most powerful when a disorder is caused by a single historical mutation. For complex traits these conditions may prove to be the exception rather than the rule.

An alternative approach to establishing a relationship between marker phenotypes and traits is to use a candidate locus. Essentially this approach simply looks for a correlation between a marker phenotype and a trait, and as such is limited to the usual argument that correlation does not imply causation. While this method is statistically very powerful, relying on mean differences between groups of differing genotypes, it is subject to various biases. Perhaps the most serious of these is the possibility that the observed association is due to population stratification. A recent development by Fulker et al[52] employs a model based on sib pair means and differences to discriminate between stratification and genuine allelic effects, this method has been implemented in Mx and applied to data on ADH2 and alcohol consumption[53]. Using sib pairs it is possible to carry out a joint test of association and linkage, which would seem to confer the advantages of both methods.

SUMMARY

This chapter has examined the merits of structural equation modeling approaches to linkage analysis. In addition to the usual advantages of maximum likelihood estimation, structural equation modeling offers a very simple approach to modeling the action of genetic and environmental factors. Four basic methods are available for univariate analysis, arising from the combination of the use of a $\hat{\pi}$ statistic or a mixture distribution with either sibling phenotypes or sibling intrapair differences. Both mixture distributions and sibling phenotype methods appear to confer additional power, with the advantage of the mixture distribution more evident when the QTL effect is large. The major strengths of the SEM approach are that it generalizes very readily to multiple loci, to complex models of the action and interaction of the environment, to multivariate, multiple rating and longitudinal cases and to models for comorbidity. When candidate loci are considered, the simultaneous modeling of means with variances and covariances allows the specification of models for joint tests of association and linkage while controlling for the effects of population stratification. Another valuable feature is that structural equation models have a simple graphical representation which facilitates the communication of complex models to audiences that are not primarily statistically sophisticated.

The author is grateful for support from PHS grants MH-40828, HL48148, MH45268, MH49492, MH41953, AA09095, RR08123, MH01458, and a grant from Gemini Holdings PLC, Cambridge, UK

REFERENCES

1 Sham, P. (1998). *Statistics in Human Genetics.* New York: John Wiley and Sons.

2 Falconer, D. S. (1989). *Introduction to Quantitative Genetics* (3rd edition). London: Longman Scientific & Technical.

3 Mather, K. & Jinks, J. L. (1982). *Biometrical genetics: The Study of Continuous Variation* (3rd edition). London: Chapman and Hall.

4 Neale, M. C. & Cardon, L. R. (1992). *Methodology for Genetic Studies of Twins and Families.* Kluwer Academic Publishers.

5 Morton, N. E. (1982). *Outline of Genetic Epidemiology.* New York: Karger.

6 Li, C. C. (1976). *First course in population genetics.* Pacific Grove CA: Boxwood Press.

7 Kruglyak, L. & Lander, E. S. (1995). Complete multipoint sib-pair analysis of qualitative and quantitative traits. *American Journal of Human Genetics, 57,* 439–454.

8 Jinks, J. L. & Fulker, D. W. (1970). Comparison of the biometrical genetical, MAVA, and classical approaches to the analysis of human behavior. *Psychological Bulletin, 73,* 311–349.

9 Eaves, L. J., Last, K. A., Young, P. A., & Martin, N. G. (1978). Model-fitting approaches to the analysis of human behavior. *Heredity, 41,* 249–320.

10 Loehlin, J. C. (1987). *Latent Variable Models.* Baltimore: Lawrence Erlbaum.

11 Everitt, B. S. (1984). *An introduction to latent variable models.* Chapman and Hall.

12 Bollen, K. A. (1989). *Structural Equations with Latent Variables.* New York: John Wiley.

13 Bentler, P. M. (1989). *EQS: Structural equations program manual.* Los Angeles: BMDP Statistical Software.

14 Jöreskog, K. G. & Sörbom, D. (1989). *LISREL 7: A Guide to the Program and Applications* (2nd edition). Chicago: SPSS, Inc.

15 Marcoulides, G. & Schumacker, E. (Eds.). (1996). *Advanced Structural Equation Modeling.* Hillsdale NJ: Lawrence Erlbaum.

16 Wright, S. (1921). Correlation and causation. *Journal of Agricultural Research, 20,* 557–585.

17 Loehlin, J. C. & Nichols, R. C. (1976). *Heredity, Environment, and Personality.* Austin: University of Texas Press.

18 Kendler, K. S., Neale, M. C., Kessler, R. C., Heath, A. C., & Eaves, L. J. (1993). A test of the equal environment assumption in twin studies of psychiatric illness. *Behavior Genetics, 23,* 21–27.

19 Hettema, J. M., Neale, M. C., & Kendler, K. S. (1995). Physical similarity and the equal environment assumption in twin studies of psychiatric disorders. *Behav. Genet., 25,* 327–335.

20 Nance, W. E. & Neale, M. C. (1989). Partitioned twin analysis: A power study. *Behavior Genetics, 19,* 143–150.

21 Neale, M. C. (1997). *Mx: Statistical Modeling* (4th edition).

22 Blangero, J. & Almasy, L. (1997). Multipoint oligoenic linkage analysis of quantitative traits. *Genetic Epidemiology, 14,* 959–964.

23 Everitt, B. S. & Hand, D. J. (1981). *Finite Mixture Distributions.* Chapman and Hall.

24 Fulker, D. W. & Cherny, S. S. (1996). An improved multipoint sib-pair analysis of quantitative traits. *Behaviour Genetics, 26,* 527–532.

25 Allison, D. B., Neale, M. C., Zannolli, R., Schork, N. J., Amos, C. I., & Blangero, J. (1999). Testing the robustness of the likelihood ratio test in a variance component quantitative trait loci (qtl) mapping procedure. *American Journal of Human Genetics.* Submitted.

26 Eaves, L. J., Neale, M. C., & Maes, H. H. (1996). Multivariate multipoint linkage analysis of quantitative trait loci. *Behavior Genetics, 26,* 519–526.

27 Vogler, G. P. (1985). Multivariate path analysis of familial resemblance. *Genetic Epidemiology, 2,* 35–53.

28 Neale, M. C., Walters, E. E., Eaves, L. J., Maes, H. M., & Kendler, K. S. (1994a). Multivariate genetic analysis of twin-parent data on fears: Mx models. *Behavior Genetics, 24,* 119–139.

29 Maes, H., Neale, M., & Eaves, L. (1997). Genetic and environmental factors in relative body weight and human adiposity. *Behavior Genetics, 27,* 325–351.

30 McArdle, J. J. & Goldsmith, H. H. (1990). Alternative common-factor models for multivariate biometric analyses. *Behavior Genetics, 20,* 569–608.

31 Kendler, K. S., Walters, E. E., Neale, M. C., Kessler, R. C., Heath, A. C. & Eaves, L. J. (1995). The structure of the genetic and environmental risk factors for six major psychiatric disorders in women: phobia, generalized anxiety disorder,

panic disorder, bulimia, major depression and alcoholism. *Archives of General Psychiatry, 52,* 374–383.

32 Achenbach, T. M., McConaughy, S. H., & Howell, C. T. (1987). Child/ adolescent behavioral and emotional problems: Implications of cross-informant correlations for situational specificity. *Psychologial Bulletin,* 101, 213–232.

33 Klein, D. N. & Riso, L. P. (1994). Psychiatric disorders: Problems of boundaries and comorbidity. In C. G. Costello (Ed.), *Basic Issues in Psychopathology,* pp. 19–66. The Guilford Press.

34 Neale, M. C. & Kendler, K. S. (1995). Models of comorbidity for multifactorial disorders. *American Journal of Human Genetics, 57,* 935–953.

35 Duffy, D. L. & Martin, N. G. (1994). Inferring the direction of causation in cross-sectional twin data: theoretical and empirical considerations. *Genetic Epidemiology, 11,* 483–502.

36 Heath, A., Neale, M., Kessler, R., Eaves, L., & Kendler, K. (1992). Evidence for genetic influences on personality from self-reports and from informant ratings. *Journal of Personality and Social Psychology, 63,* 85–96.

37 Neale, M. C., Walters, E. W., Heath, A. C., Kessler, R. C., Pérusse, D., Eaves, L. J., & Kendler, K. S. (1994). Depression and parental bonding: cause, consequence, or genetic covariance?. *Genetic Epidemiology, 11,* 503–522.

38 Snieder, H., Boomsma, D. I., Van Doornen, L. J., & Neale, M. C. (1999). Bivariate genetic analysis of fasting insulin and glucose levels. *Genetic Epidemiology, 16.* 426-446.

39 Harlow, S. D. & Linet, M. S. (1989). Agreement between questionnaire data and medical records. *American Journal of Epidemiology, 129,* 233–248.

40 Prusoff, B. A., Merikangas, K. R., & Weissman, M. M. (1988). Livetime prevalence and age of onset of psychiatric disorders: Recall 4 years later. *Journal of Psychiatric Research, 22,* 107–117.

41 Rice, J. P. & Todorov, A. A. (1994). Stability of diagnosis: application to phenotype definition. *Schizophrenia Bulletin, 20,* 185–190.

42 Rice, J. P., Endicott, J., Knesevich, M. A., & Rochberg, N. (1987). The estimation of diagnostic sensitivity using stability data: An application to major depressive disorder. *J Psychiat Res, 21,* 337–345.

43 Rice, J. P., Neuman, R. J., & Todd, R. D. (1991). Diagnostic error in linkage analysis: Application to major depression. *Psychiat Genet, 2,* 4.

44 Kendler, K. S., Neale, M. C., Kessler, R. C., Heath, A. C. & Eaves, L. J. (1993). A longitudinal twin study of 1-year prevalence of major depression in women. *Arch Gen Psychiatry, 50,* 843–852.

45 Neale, M. C. (1997). Multiple measurements increase the statistical power of genetic studies. In *Paper presented at the World Congress on Psychiatric Genetics.*

46 Kendler, K. S., Neale, M. C., Kessler, R. C., Heath, A. C., & Eaves, L. J. (1993). The lifetime history of major depression in women: reliability of diagnosis and heritability. *Arch Gen Psychiatry, 50,* 863–870.

47 Wijsman, E. M. & Amos, C. I. (1997). Genetic analysis of simulated oligogenic traits in nuclear families and extended pedigrees: Summary of GAW10 contributions. *Genetic Epidemiology, 14,* 719–735.

48 Almasy, L. & Blangero, J. (1998). Multipoint quantitative-trait linkage analysis in general pedigrees. *American Journal of Human Genetics, 62,* 1198–1211.

49 Little, R. J. A. & Rubin, D. B. (1987). *Statistical analysis with missing data.* New York: Wiley and Son.

50 MacLean, C. J., Martin, R. B., Sham, P. C., Staub, R. E., & Kendler, K. S. (1999). The trimmed-haplotype method for fine mapping of disease loci. *American Journal of Human Genetics.* Submitted.

51 Chakravarti, A. (1998). It's raining SNPs, hallelujah. *Nature Genetics, 19,* 216–7.

52 Fulker, D. W., Cherny, S. S., Sham, P. C., & Hewitt, J. K. (1999). Combined linkage and association sib pair analysis for quantitative traits. *American Journal of Human Genetics. 64,* 259–267.

53 Neale, M. C., Cherny, S. S., Sham, P. Whitfield, J., Heath, A. C., Birley, A. C., & Martin, N. G. (1999). Distinguishing population stratification from linkage disequilibrium or pleiotropy with Mx: Association of ADH2 with alcohol consumption. *Behavior Genetics.* submitted.

19

IMPLICATIONS OF PHARMACOGENETIC POLYMORPHISMS FOR HUMAN HEALTH

C. Roland Wolf
Gillian Smith

ABSTRACT

With the increasing sophistication in patient care, individuality in response to drugs in relation to drug efficacy as well as to adverse drug reactions is rapidly becoming a major issue within the Health Service. These effects are also of primary importance in the development, registration and subsequent use of new drugs. It has been known for several decades that such individuality in drug response can be genetically determined. With the dramatic input of genomics programmes and associated powerful technologies which enable rapid analysis of genetic polymorphism, there is currently great interest in establishing how such individuality affects drug development and use. The eventual goal will be the fingerprinting of individuals for drug sensitivities at an applied level. This research area, known as Pharmacogenetics or Xenogenetics, will probably be the first major area where polymorphisms between populations are characterised in detail. In order to identify and characterise these genetic differences a wide range of experimental approaches are being developed, including gene screening technologies and the study of candidate genes. Perhaps the major challenge in this research area is not the identification of novel allelic variants but the establishment of their phenotypic significance. Studies of traits in families and sib pairs may well provide a powerful approach to solving this conundrum.

ADAPTIVE RESPONSES TO ENVIRONMENTAL AGENTS

Pharmacogenetics, also known as pharmacogenomics or xenogenetics, is the study of genetic variability in response to the therapeutic or adverse effects of drugs. The genes involved in determining the pharmacological action of drugs have evolved as a fundamental part of the evolutionary process. They exist because the ability of an organism to protect itself from chemical toxins in its environment, or ingested as part of its diet, is essential for survival. When these cytoprotective mechanisms fail to cope with this environmental challenge, toxicity, cell mutation and cell death arise. Individuality in the expression of these genes is therefore of fundamental importance in protecting us from diseases where the environment is implicated. This includes almost all the major human diseases.

GENES INVOLVED IN PROTECTION AGAINST CHEMICAL AGENTS

Cellular responses to chemicals have a wide range of important implications for human health. Many protein systems determine the biological half life of therapeutic drugs, their therapeutic efficacy and potential adverse drug reactions resulting from their use. As a consequence, individuality in the levels of these proteins can be a primary concern in the drug development process, and can determine whether a drug in development succeeds or fails to reach the clinic. These genes are also of central importance in determining our sensitivity to environmental toxins and carcinogens, and inter-individual differences in their expression can result in individuality in susceptibility to disease. As these genes determine the sensitivity of cells to toxic agents, they are also important determinants of the activity of many anti-cancer drugs towards tumour cells. Indeed, the over-expression of proteins encoded by these genes can be a significant factor in drug resistance.[1]

Human response to environmental chemical agents in relation to human disease can be determined at a variety of different levels. Firstly, the circulating level of the chemical toxin or carcinogen is determined by the rate of absorption into the bloodstream. This can be determined by proteins such as the ABC drug transporters. This is a multigene family of proteins, several of which have been associated with the multi-drug resistance phenotype [2,3]. Based on predictions from the EST (Expression Sequence Tag) database, the ABC transporter gene family could include as many as twenty to thirty proteins.

Secondly, the circulating level of chemical agents is determined by the rate of hepatic metabolism by enzymes such as the cytochrome P450-dependent monooxygenases and conjugating enzymes such as the sulphotransferases and the UDP glucuronosyl- and glutathione S-transferases (see Table 19.1). If the chemical toxin reaches the target cell, the capacity to cause damage or mutations is determined by the uptake of the agent into the cell, determined in turn by the level of expression of specific drug transporters, the rate of intracellular metabolism, detoxification by enzymes such as those shown in Table 19.1 and the capacity to repair any damage which these agents induce. In this latter case, recent knowledge demonstrates that there are a large number of adaptive response systems in mammalian cells to different forms of environmental stress. These include DNA damage responses exemplified by molecules such as p53 [4,5] and also those which relate to altered cell signalling and transcriptional regulation of cytoprotective genes mediated by the MAP kinase signalling cascades.

Of these adaptive response systems, our research interest has focused on drug metabolising enzymes such as the cytochrome P450 monooxygenases, as these enzymes play a primary role in the metabolism and disposition of harmful chemical agents. There are unique opportunities associated with the study of these

Table 19.1 – Human drug metabolising enzymes

Phase 1 Functionalisation	Phase II Conjugation
Cytochrome P450	
Alcohol dehydrogenase	Glutathione S-Transferase
Aldehyde Dehydrogenase	N-Acetyl Transferase
Ketoreductase	Sulphotransferase
Quinone Reductase	UDP-Glucuronyltransferase
Flavin-containing monooxygenase	
Aldehyde oxidase	
Prostaglandin synthetase	
oxidation	conjugation with glutathione
reduction	acetylation
isomerisation	glucuronidation
hydration	sulphation
	methylation

genes in relation to genetic susceptibility to disease as, unlike most other genetic polymorphisms in man, there is a valid way to relate phenotype with genotype, i.e. the use of marker probe drugs *in vivo,* which allows us to substantiate the functional significance of any allelic variants in these genes and therefore determinine individual responsiveness to chemical agents.

THE CYTOCHROME P450-DEPENDENT MONOOXYGENASES

It has been estimated that the cytochrome P450 proteins evolved more than a thousand million years ago, initially to convert hydrocarbons into alcohols which could then be used as a source of energy. Indeed, many organisms which can live on hydrocarbons as a sole carbon source can only do so because of the presence of cytochrome P450 enzymes.[6] Since that time this enzyme system, which carries out a variety of important functions, has evolved into a multigene family of proteins in mammals. These include proteins which catalyse the biosynthesis of all steroid hormones, the metabolism of fatty acids, prostaglandins and leukotrienes, the breakdown of cholesterol to bile acids in addition to their pivotal role in the metabolism and disposition of foreign compounds. Therefore, polymorphisms in this enzyme system can have a wide variety of different consequences. In relation to cancer, for example, this can be in both the initiation and the promotion of the carcinogenic process. There are now numerous studies demonstrating the central role of the cytochrome P450 system in determining sensitivity to chemical agents, and animal studies have clearly demonstrated that the inhibition or induction of these proteins can have a profound effect on a wide range of chemical-induced toxic and mutagenic events (Figure 19.1).

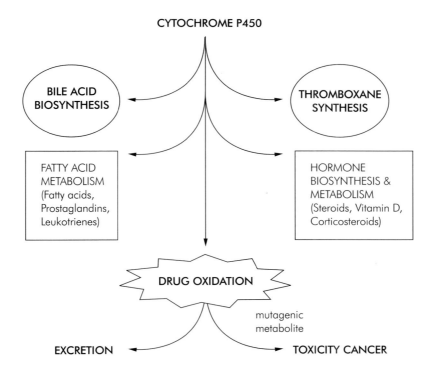

Figure 19.1 – Functions of the cytochrome P450 monooxygenase enzymes.

Ironically, in addition to its role in chemical detoxification, the cytochrome P450 system, as a consequence of the nature of the reactions which it catalyses, also has the capacity to convert chemical compounds to toxic or mutagenic products. This is the primary mechanism by which environmental carcinogens that are absorbed in the body cause cancer, i.e. they are metabolised by the P450 system to electrophilic, chemically-reactive products which bind to DNA and initiate mutations. It is also the primary mechanism of drug induced toxicities for agents such as paracetamol, which is metabolically activated in the liver to a cytotoxic product which causes damage to liver cells [7]. Individual response to chemical agents and drugs is therefore determined by the relative levels of individual cytochrome P450 isozymes, both in the liver and in the target tissue.

A Western blot demonstrating the variability in cytochrome P450 isozyme expression across a panel of human livers is shown in Figure 19.2 [8]. It is quite clear from this data that there is a very marked individuality in the expression of each of the P450 proteins. It is well known that some of this individuality is environmentally determined, i.e. that these enzymes form an adaptive response to

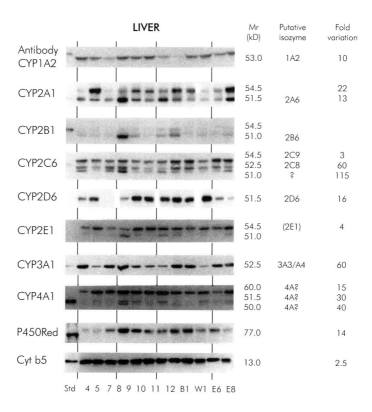

	Mr (kD)	Putative isozyme	Fold variation
Antibody CYP1A2	53.0	1A2	10
CYP2A1	54.5 / 51.5	2A6	22 / 13
CYP2B1	54.5 / 51.0	2B6	
CYP2C6	54.5 / 52.5 / 51.0	2C9 / 2C8 / ?	3 / 60 / 115
CYP2D6	51.5	2D6	16
CYP2E1	54.5 / 51.0	(2E1)	4
CYP3A1	52.5	3A3/4A	60
CYP4A1	60.0 / 51.5 / 50.0	4A? / 4A? / 4A?	15 / 30 / 40
P450Red	77.0		14
Cyt b5	13.0		2.5

Std 4 5 7 8 9 10 11 12 B1 W1 E6 E8

Figure 19.2 – Western blot demonstrating the variability in cytochrome P450 isozyme expression across a panel of human livers. Taken from Forrester et al[8]

environmental challenge and that specific hepatic isozymes are induced by a particular chemical. This can result in the increased metabolism and elimination of the chemical to which an individual is exposed. However, a certain proportion of this variability is also known to be genetically determined. We know that this genetic variability is potentially important because individuals carrying specific allelic variants of certain P450 genes can exhibit profound differences in drug disposition [9]. Perhaps the best described pharmacogenetic polymorphism in the cytochrome P450 system is that associated with cytochrome P450 CYP2D6. This polymorphism was identified following abnormal drug pharmacokinetics or adverse drug reactions from volunteers in clinical trials.[10,11] From this initial observation, these abnormalities were subsequently shown to be inherited as a familial trait, and seven or eight years ago we identified the primary gene defect responsible for this polymorphism.[12] Cytochrome P450 CYP2D6 is intriguing in that it only constitutes 1.5% of the hepatic cytochrome P450 yet is responsible for the disposition of up to 25% of all therapeutic drugs[13] (see Table 19.2, Figure 19.3). This intriguing observation could be due to the fact that this enzyme metabolises almost all CNS active agents, and could have evolved as a primary defence against neurotoxins in the environment.

Polymorphism at the cytochrome P450 CYP2D6 locus affects approximately 6% of the Caucasian population. It is, however, important to note that allelic variants at this gene locus are distributed unevenly through different ethnic groups and, in addition to gene inactivating mutations, several allelic variants which have altered substrate specificity have also been identified.[14,15] A list of some of these allelic variants is shown in Table 19.3. The importance of cytochrome P450 polymorphisms in these enzyme systems has therefore been established from pharmacogenetic studies. In addition to CYP2D6, a variety of other polymorphisms in human P450 genes have been demonstrated (Table 19.4). On the basis of the pharmacogenetic evidence, it is not unreasonable to hypothesise that genetic polymorphism in the cytochrome P450 system may be an important factor in individual susceptibility to environmentally-linked diseases such as cancer. There have now been a large number of studies carried out on this theme, particularly looking at cytochrome P450 CYP2D6.[16,17] A summary of some of our data is shown in Table 19.5. In spite of the considerable effort that has been expended on this theme, however, the role of genetic polymorphism in the cytochrome P450 system in relation to cancer susceptibility remains unclear. It is also important to note that

Table 19.2 – CYP2D6 drug substrates

β-blockers:	alprenolol, metoprolol, timolol, bufuralol, propranolol, guanoxan, indoramine, bupranolol
Anti-arrythmics:	sparteine, N-propylajmaline, propafenone, mexiletine, flecainide, encainide, procainamide
Tricyclic antidepressants:	notriptyline, desipramine, clomipramine, imipramine, amitryptiline, minaprine, fluvoxamine
Anti-psychotics:	perphenazine, thioridazine, zuclopenthixol, haloperidol, tomoxetine, paroxetine, amiflavine, methoxyphenamine, fluoxetine, levomepromazine, olanzapine, perphenazine
Analgesics:	codeine, ethylmorphine
Anti-histamines:	loratadine, promethazine
Others:	debrisoquine (anti-hypertensive), 4-hydroxy amphetamine (central nervous system stimulant), phenformin, perhexiline, MDMA (ecstacy) dextromethorphan (anti-tussive), ritonavir (HIV 1 protease inhibitor), dolasetron (anti-emetic), ondansetron, tropisetron (5-HT3 receptor antagonists), nicergoline (vasodilator), mexilitine (diabetes), dexfenfluramine (appetite suppressant), MPTP.

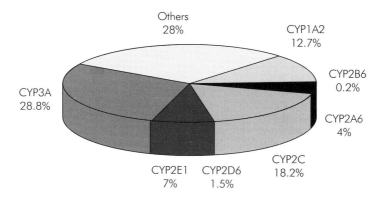

Figure 19.3A – Relative abundance of P450 isozymes in human liver.

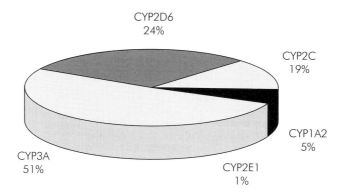

Figure 19.3B – Relative contribution of P450 isozymes to drug metabolism

meta-analysis of the data may lead to equivocal results, because it is quite possible that an allelic variant of a particular cytochrome P450 gene may be a susceptibility factor in a disease caused by one specific environmental chemical in one location, but may not be a factor in determining disease susceptibility if the same disease is caused by a different environmental agent elsewhere. In order to clarify the role of these enzymes in disease susceptibility, it is therefore important that multiple studies are carried out in the same geographical region.

OTHER DRUG AND CARCINOGEN METABOLISING ENZYMES

A number of other drug and carcinogen-metabolising enzymes are known to be subject to genetic polymorphism. Of particular interest and importance are

Table 19.3 – The major gene-inactivating CYP2D6 alleles

Allele	Trivial Name	Nucleotide substitutions	Activity
CYP2D6*1	wild-type		normal
CYP2D6*2A	CYP2D6L1		normal
	CYP2D6L2		increased
	CYP2D6L2 × 2		increased
	CYP2D62 × 12		increased
	CYP2D6N		normal
CYP2D6*3	**CYP2D6A**	**A2367 deletion**	**INACTIVE**
CYP2D6*4	**CYP2D6B**	**G1934A**	**INACTIVE**
CYP2D6*5	**CYP2D6D**	**gene deletion**	**INACTIVE**
CYP2D6*6	CYP2D6T		inactive
CYP2D6*7	CYP2D6E		inactive
CYP2D6*8	CYP2D6G		inactive
CYP2D6*9	CYP2D6C		reduced
CYP2D6*10	CYP2D6J		reduced
	CYP2D6Ch		reduced
	CYP2D6M		reduced
CYP2D6*11	CYP2D6F		inactive

Table 19.4: Polymorphic human P450s

Enzyme	Chromosomal localisation	No. of alleles	Major alleles
CYP1A1	15q 22-qter	4	*Msp* I RFLP 3' non-coding $T_{5639}C$ 3' non-coding I_{462} V, Exon 7
CYP1B1	2p22-21	6	Exon 2 $Arg_{48}Gly$ Exon 2 $Ala_{119}Ser$ Exon 3$Val_{432}Leu$ $Asn_{453}Ser$
CYP2A6	19q13.1-13.3	3	CYP2A6*2, T->A Exon 3 CYP2A6*3, C-> A Exon 3
CYP2C9	10q 24.1-24.3	3	CYP2C9*2, C-> T Exon 3 CYP2C9*3, A->C Exon 7
CYP2C19	10q 24.1-24.3	10	CYP2C19*2, G->A Exon 5 CYP2C19*3, G->A Exon 4 CYP2C19*4 initiation ATG->GTG CYP2C19*5 Arg_{433} Tryp CYP2C19*6 $Arg_{132}Gln$ CYP2C19*7 GT->GA intron 5 CYP2C19*8 $Tryp_{120}Arg$
CYP2D6	22q 11.2-qter	>18	CYP2D6*3, del A_{2637}, CYP2D6*4, G_{1934A} CYP2D6*5, gene deletion

Table 19.5 – CYP2D6 genotype frequencies in cancer patients and controls

Sample	Sample size		Genotype[1] (%)			p-value
	(n)	EM	HEM	PM		
Control	720	66.1	29.6	4.3		
Emphysema	151	61.6	35.1	3.3		
All cancers	1759	63.4	31.6	5.0		0.05*
Lung cancer	361	64.8	31.6	3.6		
Breast cancer	437	66.8	29.3	3.9		
Colon cancer	115	63.8	29.6	6.9		
Leukaemia	312	65.7	26.9	7.4		0.02*
Teratoma	169	65.7	30.8	3.5		
Melanoma	127	54.3	39.4	6.3		0.02*
Bladder cancer	184	53.8	41.8	4.4		0.005*
Prostate cancer	54	59.2	31.5	9.3		
Parkinson's disease	229	60.7	27.5	11.8		0.0001*

[1]CYP2D6 genotype was investigated by determining the presence of the CYP2D6*3, CYP2D6*4 and CYP2D6*5 alleles; EM - extensive metaboliser, HEM heterozygous extensive metaboliser, PM poor metaboliser;* statistically significant

polymorphisms at the glutathione S-transferase and N-acetyl transferase gene loci (Table 19.6). Both of these multigene families of proteins have the capacity to activate or predominantly inactivate carcinogenic products generated by the cytochrome P450 system. Modulation of the level of these proteins in laboratory animals can also have a profound effect on chemical toxicity and carcinogenicity. Indeed, we have recently deleted one particular glutathione S-transferase pi gene from the mouse genome and demonstrated that Gstp1 null animals become hypersensitive to the carcinogenic effects of polycyclic aromatic hydrocarbons, similar to those found in cigarette smoke [18]. Intriguingly, glutathione S-transferase pi is also polymorphic within the human population, with allelic variants demonstrating significant differences in their substrate specificity[19,20]. Therefore, these allelic variants may be of importance in the aetiology of cancers where exposure to cigarette smoke is thought to play an important role. Some of our data on this theme is summarised in Table 19.7. Whereas we found some evidence that the GSTP1 polymorphism may be a susceptibility factor in bladder and testicular cancer, we also found that there was a shift in allele frequency in lung cancer patients relative to controls[19]. However, this was not significant. In a subsequent study, however, a significant difference in allele frequencies between control and lung cancer populations was established [21]. This data therefore allows the extrapolation of the role of this gene in susceptibility to tumorigenesis in animal models to the epidemiological evidence obtained from a study in man. However, these data also require substantiation.

Genetic polymorphisms, which result in individuality to the beneficial or adverse effects of therapeutic drugs can be of great economic and clinical importance. The

Table 19.6 – Polymorphisms in human GST and NAT genes

A: Glutathione S-transferases[1]

Gene	Allele	Sequence changes
GSTM1-1	GSTM1null	gene deletion
	GSTM1a	Exon 7, 173Lys
	GSTM1b	Exon 7, 173Asn
GSTT1-1	GSTT1 null	gene deletion
GSTP-1	GSTP1a	consensus
	GSTP1b	$Ile_{105} - > Val$
	GSTP1c	$Ala_{114} - > Val$

B: N-Acetyl transferases[2]

Gene	Allele	Sequence changes
NAT1	NAT1*4	consensus
	NAT1*10	$T_{1088}A$
	NAT1*11	$Val_{149}Ile$
	NAT1*14	$Arg_{187}Gln$
	NAT1*15	$Arg_{187}STOP$
	NAT1*17	$Arg_{64}Tryp$
	NAT1*19	$Arg_{33}STOP$
	NAT1*21	$Met_{205}Val$
	NAT1*22	$Asp_{251}Val$
	NAT1*25	$Ile_{263}Val$
NAT2	NAT2*4	consensus
	NAT2*5A	$Ile_{114}Thr$
	NAT2*6A	$Arg_{197}Gln$
	NAT2*7A	$Gly_{286}Gln$
	NAT2*14A	$Arg_{64}Gln$
	NAT2*17	$Glu_{145}Pro$
	NAT2*18	$Lys_{282}Thr$

[1] Nomenclature according to Mannervik et al[22], [2]Nomenclature according to Vatsis et al[23]. Only allelic variants which lead to a change in amino acid sequence are shown. Further alleles have been identified which contain single nucleotide polymorphisms – the phenotypic consequences of these are presently unclear.

identification of genes involved in such pharmacogenetic responses has predominantly relied on taking a clinical observation of an adverse drug reaction and identifying the gene responsible for the metabolism of the drug concerned or the receptor with which it interacts. There have been very few studies to try to identify novel gene loci using classical genetic approaches such as family studies or sib pair analysis. One advantage of the study of pharmacogenetics is that populations can be given low doses of drugs and any aberrant phenotype, for example, metabolism can be determined. Both family studies and sib pair analysis could provide a powerful approach to identifying novel pharmacogenetic polymorphisms. However, in

Table 19.7 – GSTP1 genotype frequencies in cancer patients and controls

Sample	Sample size (n)	GSTP1a/1a	Genotype GSTP1a/1b	(%) GSTP1b/1b	Odds Ratio (95% CI)	p-value
Control	155	51	42.5	6.5		
COPD	79	43	44.3	12.7	2.1 (0.8–5.8)	0.174
Lung cancer	115	43.5	45.2	11.3	1.9 (0.7–4.8)	0.233
Testicular cancer	155	38.7	42.6	18.7	3.3 (1.5–7.7)	0.002*
Bladder cancer	40	37.5	40	22.5	4.2 (1.4–12.4)	0.005*
Prostate cancer	36	27.8	69.4	2.8	0.4 (0.02–3.3)	0.693
Breast cancer	62	40.3	51.6	8.1	1.3 (0.4–4.3)	0.77
Colon cancer	100	37	55	8	1.3 (0.4–3.6)	0.825

COPD – chronic obstructive pulmonary disease; * statistically significant

order for these approaches to be applied, phenotyping analysis would need to be carried out on the groups of individuals to be studied. Logistically this could be quite difficult to do.

CONCLUSION

A wide number of genes have evolved which protect cells from the deleterious effects of environmental chemicals. Many of these genes in man are polymorphic and therefore are good candidate genes to establish whether individuality in their expression may relate to individual susceptibility to cancer. However, these are only a few of a large number of genes that may also be involved in determining individuality in disease susceptibility. Some of these genes have already been identified as potential candidate genes, whereas there clearly will be others which may only be identified by applying more sophisticated classical genetic linkage studies.

REFERENCES

1 Hayes JD and Wolf CR. Molecular mechanisms of drug resistance. *Biochem. J.* 1990; **272**; 281–295

2 Gottesman MM, Hrycyna CA, Germann UA and Pastan I. Genetic analysis of the multidrug transporter. *Ann. Rev. Genet.* 1995; **29**; 607–649

3 Borst P and Schinkel AH. Genetic dissection of the function of mammalian P-glycoproteins. *TIG* 1997; **13**; 217–222

4 Hollstein M, Sidransky, Vogelstein B and Harris CC. p53 mutations in human cancers. *Science* 1991; **253**; 49–53

5 Harris CC. p53: at the crossroads of molecular carcinogenesis and risk assessment. *Science* 1993; **262**; 1980–1981

6 Nebert DW. Multiple forms of inducible drug-metabolizing enzymes: a reasonable mechanism by which any organism can cope with adversity. *Mol. Cell. Biochem.* 1979; **27**; 27–46

7 Zhou L, Erickson RR, Hardwick JP, Park SS, Wrighton SA and Holtzman JL. Catalysis of the cysteine conjugation and protein binding of acetaminophen by microsomes from a human lymphoblast line transfected with the cDNAs of various forms of human cytochrome P450. *J. Pharm. Exp.Ther.* 1997; **281**; 785–790

8 Forrester LM, Henderson CJ, Glancey MG et al. Relative expression of cytochrome P450 isoenzymes in human liver and association with the metabolism of drugs and xenobiotics. *Biochem. J.* 1992; **281**; 359–368

9 Tucker GT. Clinical implications of genetic polymorphisms in drug metabolism. *J. Pharmacol.* 1994; **46**; 417–424

10 Mahgoub A, Idle JR, Dring LG, Lancaster R and Smith RL. Polymorphic hydroxylation of debrisoquine in man. *Lancet* 1977; **2**; 584–586

11 Eichelbaum M, Spannbrucker N, Steinke B and Dengler HJ. Defective N-oxidation of sparteine in man – a new pharmacogenetic defect. *Eur. J. Clin. Pharmacol.* 1979; **16**; 183–187

12 Gough AC, Miles JS, Spurr NK, Moss JE, Gaedigk A, Eichelbaum M and Wolf CR. Identification of the primary gene defect at the cytochrome P450 CYP2D locus. *Nature* 1990; **347**; 773–776

13 Shimada T, Yamazaki H, Mimura M, Inui Y and Guengerich FP. Interindividual variations in human liver cytochrome P450 enzymes involved in the oxidation of drugs, carcinogens and toxic chemicals: Studies with liver microsomes of 30 Japanese and 30 Caucasians. *J. Pharmacol. Exp.Ther.* 1994; **270**; 414–423

14 Smith G, Stanley LA, Sim E, Strange RC and Wolf CR. Metabolic polymorphisms and cancer susceptibility. Cancer Surveys 1995; **25**; 27–65

15 Daly AK, Brockmöller J, Broly F et al. Nomenclature for human CYP2D6 Brockmöller alleles. *Pharmacogenetics* 1996; **6**; 193–201

16 Wolf CR, Smith CAD and Forman D. Metabolic polymorphisms in carcinogen metabolising enzymes and cancer susceptibility. *Br. Med. Bulletin* 1994; **50**; 718–731

17 Rostami-Hodjegan A, Lennard MS, Woods HF and Tucker GT. Meta-analysis of studies of the CYP2D6 polymorphism in relation to lung cancer and Parkinson's disease. *Pharmacogenetics* 1998; **8**; 227–238

18 Henderson CJ, Smith AG, Ure J, Brown K, Bacon EJ and Wolf CR. Increased skin tumorigenesis in mice lacking pi class glutathione S-transferase. *Proc. Natl. Acad. Sci.USA* 1998; **95**; 5275–5280

19 Harries LW, Stubbins MJ, Forman D, Howard GCW and Wolf CR. Identification of genetic polymorphisms at the glutathione S-transferase Pi locus and asociation with susceptibility to bladder, testicular and prostate cancer. *Carcinogenesis* 1997; **18**; 641–644

20 Matthias C, Bockmühl, Jahnke V et al. The glutathione S-transferase GSTP1 polymorphism: effects on susceptibility to oral/pharyngeal and laryngeal carcinomas. *Pharmacogenetics* 1998; **8**; 1–6

21 Ryberg D, Skuag V, Hewer A et al. Genotypes of glutathione transferase M1 and P1 and their significance for lung DNA adduct levels and cancer risk. *Carcinogenesis* 1997; **18**; 1285–1289

22 Mannervik B, Awasthi Y, Board PG et al. Nomenclature for human glutathione transferases. *Biochem. J.* 1992; **282**; 305–306

23 Vatsis KP, Weber WW, Bell DA et al. Nomenclature for N-acetyltransferases. *Pharmacogenetics* 1995; **5**; 1–17

INDEX